Johannes Goedaert

Johannes Godartius Of insects

Johannes Goedaert

Johannes Godartius Of insects

ISBN/EAN: 9783744670913

Printed in Europe, USA, Canada, Australia, Japan

Cover: Foto ©berggeist007 / pixelio.de

More available books at **www.hansebooks.com**

Johannes Godartius

O F

INSECTS.

Done into Englifh, and Methodized, with
the Addition of Notes.

The Figures *Etched upon* Copper, *by* Mr. F. Pl.

Y O R K.
Printed by *John White,* for M. L. 1682.

To the Reader.

I Am not satisfied, that ever Goedartius intended to publish these Papers: or that He aimed at any thing more, then to please and gratifie his Fancy, and the excellent skill he had in Limning. And this appeares from the almost totall neglect of descriptions, which he had sufficient oppertunity, to have well performed; and the many references to the Colours in his Designs, which Calcography, or Printing could not represent; besides the tumultuous order, we have them first published in, after his Death: above all, the little advancement he seemes to have made, in his Skill, in the nature of Insects, after 40. Yeares (as he says himselfe) daily conversing with them: so that he seemes rather, to have diverted with them, then to have given himselfe the trouble of well understanding them.

And yet after all this, you will find him every where very just, and true in his Observations: but in many places very short, and hardly Intelligible.

Again, as he most industreously fed them, and

A 3   observed

observed their Changes ; *so he committed little
or nothing to Writing or* Designe, *but what succeeded with him, and (as he understood it)
had its right* Change : *Which is more, than any
man ever did before him* ; *So that we need not
admire, that so long, and pertinatious an Industry
produced so few* Historys : *For he* Designed *not
all, that came to hand, but such only, as it was his
good fortune to Feed, and bring up to* Change.

  And yet in these few Historys, *you will have
something of all the severall* Genus's *of* Insects *,
that are in Nature.*

  He left his Papers in Dutch, *which were at
severall times, and that by divers hands, put forth
in* Latin : *His* Latin Interpreters *have added*
Comments, *indeed* ; *but were Men wholy ignorant
in* Naturall History; *and their Comments are meere*
Rapsodies, *and altogether impertinent to the Ex-
plication of any one* History *of* Godartius : *As he,
who shall attentively consider them, will plainly
find : wherefore in this our* Translation; *I have o-
mitted them, and have Printed the* Authors *words
only, and have here and there used the freedom of
animadverting, upon what is not agreeable to my
owne experience in these matters : besids I have*
Methodised *the* Historys, *according to the Severall natures of the* Insects *they treat about* ; *and;*
                                                where

To the Reader.

(where I could underſtand him) I have explained many things, that might be doubted of: And thoſe who injoy a ſutable leaſure may hereby be put in a ready way, to purſue the History *of theſe long neglected* Animals.

Alſo I have taken great care of the Deſignes, *in transferring them upon* Copper Plates; *which I dare promiſe are* Exquiſitly *performed, by the beſt of our* Engliſh Artiſts: *which was my expence; and which the Book-ſellers were not willing to Reimburſe me; ſo that this Impreſſion conſiſts but of a* 150. Coppys, *which were intended only for the curious. And upon this occaſion I muſt needs ſay, that* Naturall Hiſtory *is much injured, through the little incouragement, which is given to the* Artiſt, *whoſe* Noble *performances can never be enough rewarded; being not only neceſſary, but the very beauty, and life of this kind of learning.*

If the Reader *ſhall yet deſire to be informed, why the* Comments *go no farther, then the greater part of the work only; let him know, that the* Tranſlation *lay by me unfiniſhed, and in the firſt draught for above* Seven Years, *and had ſo done long enough, had not my very good Friend* Mr. T. K. *of* Cowkreig, *very* Obligingly *undertaken to* Tranſcribe *and perfect it. So that the humour of Commenting*

ing being past me, and the notions I once had of these things, being lesse fresh in my memory, I had little leisure to instruct my selfe againe, but suffered it to go in that particular, as he found it.

M. L.

Section I.

Of Butterflyes, *fitting with erect Wings:* all whofe Chryfolis are Angular: of which there are VIII. Species.

Number. 1.

THE Catterpillar, *N.* 1. is prickle Haired; it hath its birth from an Egg, which the Butterfly (*N.* 1.) haveing eyes in its Wings, like thofe in Peacocks feathers, doth lye upon Netles.

This Catterpiller I took up the 14 day of *May*, 1635. And I fed it with the Leaves of Netles, untill the 11. of *June*, of the fame Year. Then it compofed it feife for change , with its head hanging downwards, as you may fee in the Table. It remaind in this forme 19 dayes. When a moft elegant Butterfly came forth.

At the firft coming forth of the Butterfly, its wings were like wet Paper ; Off of which fell certain watery dropps: But that, which did feem to me worthy Obfervation, they became in halfe an houre dry, expanded, and fit for flight.

This Butterfly feeds on fweet things, as Sugar, and the Honey of flowers: Alfo it is mainly delighted in rotten fruits, for which they fight among them felves.

B   In

In winter-time they hide them selves within the Chimnies of poor Cottages, from whence I have forc't them with a good blazing fire.

Also they are found lying hid in hollow Trees.

It is well Observed by our Author that this Catterpillar, hath its beginning from the Egg of such a Butterfly ; and so probably have all Catterpillars whatsoever their beginning from the Eggs of their respective Butterflys. The Butterfly is the Mother Insect in perfection, and the Catterpillar, its Aurelia, or Chrysalis are but certain Disguises for a time, wherewith one and thesame Animal is by nature invested for divers ends. viz, that of the Catterpillar to eat such and such food ; This of the Aurelia to perfect and harden its limbs.

Number. 2.

Godar.
Part 1.
Tab. 12.

Although our Fingers suffer and are stung by Nettels, yet the Catterpillar marked N. 2. delightes to feed on them: Neither doth it make ready for the change, or obstaine from food, whilst this Plant flourishes.

It began to change the 23. of *June*, and the 9th. of *July* came forth the Butterfly marked with beautifull colours, *Figured* in N. 2.

These Butterflyes are to be found all Winter in Stables, where Beasts stand.

These Catterpillars are exceeding Voratious.

These

Here is wanting the Figure of the Chryſolis. We have noted in the foregoing Number, the uſe of the diſguiſes of Butterflyes: and I ſhall here Obſerve, that Catterpillers feed of courſe and barſh food, ſuch as the ſubſtance of the leaves of Plants; whereas the Butterſtyes feed of the Honey of Flowers, and liquid meats. This is contrary to what is naturall in ſangvineous and more perfect Animals; who in the Embryo feed of a prepared Chyle, but after birth have a yet more courſer food to nouriſh them: And yet more courſe as they grow older, and to maturity.

Number. 3.

The nouriſhment of the Catterpillar *N. 3.* are the leaves of the *Elme.*

When the time of the change grows near, they betake themſelves to Houſes, and fix their hinder parts to a wall, hanging down with their heads, that they may more eaſily come forth of their ſhell or *chryſalis,* when the time of change is compleated.

Before the Catterpillar changes its ſhape for that of a *chryſalis,* and puts off its old skin, it ſeems to be very much troubled, turneth, ſhaking, and toſſing its body every way, and trembling as if it had an Ague. At length riſing, and falling often with his body it conducts its body into a circle, upon which it ſwells ſo, that the skin cracks all the length, and ſo by little and little it falls off, a new skin growing underneath. And at that time they reſt a while.

B 2   This

This is very notable in these Catterpillars, that where the back of the Catterpillar was, there are the belly and feet of that Animal it's changed into ; and the contrary, where the belly and feet of the Catterpillar were, there now the back of that Animal is, which was produced by the change of the Catterpillar. And this change is produced in a very short time ; so that it may diſtinctly be ſeen and obſervcd : For as ſoon as the old skin is layed aſide, this Transfiguration manifeſtly appears.

The Catterpillar of *Number* 3. Began to change the 21 ſt. of *June* and the 30th. day of the ſame Month it was changed into a very fair Butterfly, which I have alſo exactly delinated *Number,* 2.

Immediatly after the change, its wings were like moiſt Paper, but in halfe a quarter of an hower they dryed, were expended and made fit for Flight.

Afterwards the Butterfly let fall from his Taile four dropps of blood ; and after halfe an hower one drop of cleer liquor, like Fountain Water.

It lived a Butterfly 19 dayes without food.

Another Catterpiller of the ſame *Species* compoſed it ſelfe for Change the 13th. of *July*, and that change being come to perfection. there broke out of the Back, (of the Catterpillar) 82 Flyes, as in *Number* 3. So that out of one and the ſame *Species* of Catterpillars, a Butter fly is produced, and 82 Flyes.

Note, *That I believe the Author was miſtaken in his Obſervation of the Back of the Catterpillar being the Belly of Butterfly, and the* Chriſolis *might well enough I conceive turn its body within a dry, and looſe skin, in the act of throughing it off, and cauſe this miſtake ; eſpecially, when it was ſo much troubled and concerned in the cracking of it, as is noted by our Author.*

2. Con-

2. Concerning the Catterpillars fixing his Body to a Wall, it is to be Noted, that this is done by a fingle thred crofs the midle, thus, (for I have, more then once actually feen it in doing) This is done, before it appeares in the difguife of an angular *chryfolis*. The Catterpillar doubles its head backwards, and touching the place, where it would fufpend it felfe, it fixes a thred on both fides its body, drawing it a croffe, and then reducing its head and laying it felfe in a pendulous pofture, it toffes it felfe and cracks the skin of the Catterpillar, which flying off, it appeares a *chryfolis*, hanging as is defcribed.

3. I have my felfe often Obferved that Red liquor fall from the Butterfly *Number*, 2. And alfo an equivalent one, though not Red, from many flyes hatching from *chryfalifes* : and 'tis by the meanes of this liquor, they fwell their Bodies and crack their Shells of their refpective *chryfalifes*. A fprinkling alfo of this liquor makes their wings fo moift , when they firft come forth. See in the Life of *Peireflius* of the Reining of Blood, referred hither.

4. the 82 Flyes here mentioned were the brood of the *Ichneumon Fly*, conveied into the body of the Catterpillar. I confefs I am not yet acquainted with the manner of the conveying them ; but I am perfwaded that it is done by the *Ichneumon parent*. This I can affirm, that thefe bold Animals do frequently lay their Yong in the very Egg-cake of Spiders. Again I have feen them perforating the excrefcencies of trees with a tongue from their mouth, as with a winble ; without doubt either to deftroy and feed on the Maggot within the excrefcence ; or elfe to impregnat them with their owne kind. More I have fpoken to this matter in one of the monthly Tranfactions of Mr. *Oldenbourgh* : However thefe Flys moft certainly are a *By-birth* only, It is farther remarkable in this Number, that thefe 82 Flyes did break forth of the *chryfalis*, which alfo is unufuall.

Number

Number 4.

Some times I have obſerved the Changes of Inſects to be made into more beautifull Animals ; then I expected ; as in the Catterpillar *N.* 4.

I fed him with *Elme* leaves, in which he delighted. He hath a very deformed and ugly head, whereas in in moſt other Catterpillers the head ſhineth like a Looking Glaſs.

He purgeth himſelfe and beginneth to change into the forme *N.* 4. The 9th. of *June,* and the 20th. of the ſame Month came forth a moſt beautifull Butterfly, marked with moſt elegant colours.

Theſe Butterflyes continue alive all Winter, unleſſe other little Animals devour them, or Spiders ſtrangle them in their webs.

Our Author hath diligently obſerved all alonge, that Catterpillars exactly Purge themſelves of all their excrements before they change, and are in the diſguiſe of Chryſaliſes. *We have ſaid that the uſe of the diſguiſe of a Catterpillar is, to eat a different Food, and that which the Butterfly cannot eat : And therefore it is but reaſonable , that the Catterpiller ſhould quite and cleane empty it ſelfe of all the old food, when it is about to become a new feeder. I am moreover conceited, that the Change of a Catterpillar is not ſuperficial only, but goes deeper yet, and that the inteſtines are in ſome ſort changed alſo, as well as the Organes of the mouth. The inſide of the Gutts being indeed an outſide too in all Animals.*

Number. 5.

I gave this Catterpillar *N.* 5. Many plants to eat but he refused them all: At length it came into my minde, that he might be delighted with Netles, which when I had brought to him, he to my great admiration rubd his Head against them, and shewed signes of gladnefs, and eat them greedily.

After he had been fed some time with Netles, he composed himselfe for change *October,* 3. In the forme expreſſed *N.* 3. Under a Glass, to which he fixed himselfe with his Head downwards.

If you touched the *chryfalys,* it moued it selfe so ftrongly as made the Glaſſe ring like a Bell.

Out of this change came the 1ft. of *December,* a very faire Butterfly, Peacock-like Eyed.

The Butterfly lived 40 Dayes without food, and dyed, for I knew not what to feed him with.

This Catterpiller ſayes 'our Author, refuſed to eat of many forts of plants offered him, and would feed of none but Netles: I add, and he would have undoubtedly ſtarved firſt:This is not an effeⱪ of the different make of the Organ of their mouthes, (as above ſaid betwixt the feeding of a Catterpillar and a Butterfly) but verily the delicatenſs of the Palate and taſte : which perhaps might be improved to good purpoſe in diſcovering the vertue of Plants.

I am perswaded there is no better way to know what kind of wood is best, for Sheathing of Ships, than to essay certain pollished piecesthereof like Tallies tyed to a Buoy In tho waters and streames much infested by the Worme. For that sort of wood, which they shall refuse, is in all reason to be chosen for the use desired. And the Indies are stored with greater variety of Timber, then Europe, so that it would be very probable there may he some found, which that kind of River Worme will absolutely refuse to Eat.

O.P. 3.
Tab. 4.

Number. 6.

The Catterpillar marked *N. 6.* Leads a peacefull Life : But is exceeding fearfull, it is moreover hairy, and very sensible,

It feeds on the leaves of *Carduus Benidictus* the softer part of which leaves they greedily eat; but the more sticky and harde nerves they do not touch.

About evening they make themselves Webbes in which they defend themselves against the cold of the night. They are of a Black colour, and have sharpe pricles mixt with yellow.

This Catterpillar began to change the 19th. of *July* 1665. The change was of a pretty Figure, and as it were Guilt.

Out of the *Chrysalis* came the 7th. of *August*, a most elegant Butterfly and marked with divers conolrs. It lived with out Food, untill the 11th. of the same month.

The

(9)

The midle state or disguise of a Butterfly is called by the Greeks Chrysalis, or a thing guilt, as the word importeth : The Latin hath left us no name, that I know off : we have Translated it Aurelia. The Latines however call the Catterpillar Eruca : which is a word (as I guesse from a place in Vitruvius) which signifies in the old Tuscan Language Viride æris, and thence borrowed to signifie a Catterpillar ; for some Catterpillers there are, which I have seen in Languedoc feeding on a certain Common Tithymal, very notably painted with a sea green Colour, or Blew. So that as the guilding of some few Chrysalifes, gave a denomination to all ; in like manner the Blew Colour of some one Catterpillar gave the Name to all the rest. As for the Guilding it selfe , I take it to be nothing elfe , but the Scum of an evaporated juice between the Skin of the Catterpillar, and the shell of the Chrysalis, which last it covers.

Number. 7.

G.P.1.
Tab. 11.

Few Catterpillars love Cabbage, and yet the Catterpillars which are designed in the 7th. Table : eat white Cabbage, but will not touch the Red Cabbage. Cold and moist weather is a very great enemy to them and soon destroys them , and they wither to nothing, but skin.

They have a double time of change ; if that happen in Summer, it is at an end soone, but if it begin in Autumne, it lasts untill the following Summer ; I have experienc'd both changes.

One of the Catterpillars of the 7th. Table changed the First of July, and the 12th. of August, the white

C   Butterfly

Butterfly came forth, reprefented in this Table ; Another of the fame fpecies, changed the 17th. of *December* and remained a *Chryfalys*, till the 15th. of *May* the Year following ; when a Butterfly came forth ; theyery fame with the former.

But another Year it happened, that I obferved in the fame Catterpillars a wonderfull thing, I tooke a certain number of thefe Catterpillars, at the fame time : I fed them untill they of their own accord left their meat, and betook themfelves to reft and for generation : after they had lain ftill 4 dayes,and did not move,I faw break forth of the Skins of each Catterpillar, on both fides the Animall 40. in fome, 50. in fome 52 little Wormes, which Wormes as foon as borne, made themfelves little Netts : (or Baggs) of yellow Silk, beginning from the Taile to the Head, and fhut themfelves up in thole Netts in thofe Baggs they defend themfelves from the cold of Winter.

The Catterpillar(out of whofe Skin I faid thofe Worms came) knit all their little Netts together on a bunch that they might not be fcattered, but that they might be turned into Flyes in Summer, in one place, and at once.The Catterpillar notwithftanding all thefe wounds, out of which 40, or 50 Wormes did break forth, lived without Food in my Clofíet from the 24th. of *September*, untill the 28th. of the fame Month. the 19th. of *October* the above defcribed Worms turned into fo many little Flyes, and all of them dyed within 6 Dayes.

Another Catterpillar of the fame Species after its Change, and that it hadlaine in it 14 Dayes, 2 Worms broke out of the forehead of it, and thofe two Wormes, in my fight in the fpace of an hour and a halfe, were changed into Eggs of an Amber colour : and 13 Dayes after that, out of each Egge came a middle fifed Fly.

Thefe things I have had the experience of, and have Obferved them, not without admiration, becaufe it seems besides, if not against the usuall course of Nature, that from one and the same Species of Animals, an Offspring of different species shou'd be gendred, and that one and the same Animal shou'd procreate after divers manners, which thing yet is made manifest in these Catterpillers, as I have in few Words declared.

A double time of Change] The reason may be, for that those Catterpillars, as is Observed by our Author, are very tender; so that they which Change not, till the cold Weather come on, continue in the middle state or disguise, till the Spring following: otherwise shou'd they then Change into Butterflyes and lay their Eggs, the Brood would perish with the Cold: Others there are to my knowledge, that are constantly hatch'd Catterpillars the latter end of Summer, and not having time and Food to bring them up to a full growth, in order to their Change, do club for a web amongst themselves, and continue a very small Fry all Winter, and when the leaves begin to break forth, they again come forth, feed a new, grow great ones, and Change: as the common Hedg Catterpillar, &c.

2. The numerous Rase are Ichneumon's, the two others are a sort of Flesh-fly: Both these I say are By-births, and not at all generated by the Catterpillar, but by their respective Parents: the Catterpillar which bore them, serving only as Food to them, not a Mother. It's to be Observed that the Flesh-Flyes, did feed upon the very substance of the Catterpillar, or Chrysalis, as they would upon Carrion. That the Ichnenmons did not destroy the Mother, and love not corrupt meat, possibly the very food of the Catterpillar digested by her, was their nourish-

ment

*ment, and not her bowels, who many days survived the
strainge eruption of that brood; I have opened many Catterpillers of that very Species, in which I have found of these
Worms, but how and when they are conveyed into their Bodies, I do not yet understand.*

*3. That is very curious and particular, if there be no
mistake in the thing, that our Author sayes : The Catterpillar her selfe knit all the little Netts into one bunch, as though
she acknowledged this Brood to be her owne : and yet this is
no proofe they are so ; For we see in Birds, the like instance,
the little Bird called the* Hedg Sparrow, *will carefully and
most affectionatly, bring up, as well as Hatch, the young* Cuck-
ow.

Number 8.

When Savoy Cabbages and Colliflou'rs begin to
knit, often are seen about them certain Butterflyes
which lay their Eggs upon those Plants, which Eggs
by the heat of the Sun, are Attracted and set in an up-
right Posture, and at length Catterpillars break forth
of those Eggs and devower the Plants.

These Catterpillers easily endure the heat of the Sun,
but not at all lasting Raines, which soone maketh them
tabid and nothing but Skin.

The Catterpillar of this 8th. Table, purged itselfe and
changed the 3d. of *August*, and the 17th. of the same
Month came forth the Butterfly here represented ; This
Butterfly is but slow and not nimble, though it be sometimes found to live over the winter and longer.

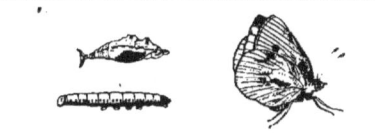

*I have verily Obſerved the Butterfly in the very act
of laying her Eggs upon a cabbidg leaſe; which is thus done,
the Butterfly ſitts upon the edge of the leaſe, and bending her
tayle under the leaſe. ſhe fixes (by touching the leaſe with
the point of her taile) one Egge, and then an other, and
ſo a 3d. untill ſhe hath layd as many as ſhe liſt ; But all
theſe Eggs are in that moment fixed with the ſmall end upwards, and not laid ſidewayes, and afterward turned upwards, by the heat of the Sun, as our Author ſeems to ſay.*

Section. II.

Of Butterflyes ſitting with flat Wings, whoſe *Catterpillers* want the middle Leggs, and from there manner of going in *Loops*, are by ſome called *Geometræ.*

Number. 9.

G. P. 1.
Tab. 31.

The Catterpillar of the 9th. Table comes of Seed
or Eggs, hard to be found out or met with, I at length
with much dilligence Obſerved them, laid upon certain leaves, and covered with haire and downe, Animall
like, and ſo are preſerved ſafe from cold, opening thoſe
leaves I found a Green ſeed, or Eggs.

Theſe Catterpillars feed on *Gooſebury*-leaves, alſo of
Red and White *Curran-berries.*

They are wont to change about the end of *June*, as
the 22. The change is elegant, whoſe Figure you have

C 3   in

in this Table. It remained in the chang to the 13ᵗʰ· of *July*, on which day came forth a small white Butterfly with black spots.

If you take the *Moth* in your hands, or suffer it to fall, it seems dead a while and moves not,

Here is a great mistake in this History, *of Catterpillars laying of Eggs, and therefore I have amended it in the very* Text, *beliving it an escape of the* Latin *Translation.*

This Catterpillar feeds not only upon the leaves, but on the very green Fruit *of the* Apricock *, as I have Obscrved.*

It was from the change of one of these Catterpillars that I imagined I had a small green Beetle, *but more of this in another place.*

G.P. 2.
Tab. 34.

Number. 10.

The Catterpillar of the 10ᵗʰ· Table : was put into my hands by a couple of friends, the 14ᵗʰ· day of *October*, with intreaty that I would observe its way of liveing and change, I undertook to do what lay in my power, I knew not what it fed on, and therefore I took much paines to find it out, to that end I put before it leaves of divers Herbs and other things, which I knew Catterpillars to feed off, but it would not so much as taste of any of them. Then from its colour, which was not much unlike the leaves of *Elder*, I began to suspect, that those leaves might not be ungratefull to it, being by experience taught, that some Catterpillars, are coloured some what like the Herbs and leaves they are nourished with : I gave it *Elder* leaves to eat, it devowred those willingly as I thought, but yet once a day only.

At evening after Sunset, when it began to be darke, it eat halfe an hour together, when it once fell on, and being full it stretched it selfe right out, and in that posture it looked like a stick out of a Faggot, for I cou'd see no leggs at first look, and yet was not without them, wherefore I have drawn the feet of the forepart of the Body, in that part also was a rising bunch ; by the halfe of which it so fastned it selfe, sucking continually like a *Leech*, that it cou'd scarce be plucked away, and wo'd rather suffer it selfe to de plucked in pieces, then be separated from what it stuck to.

Also the hinder part of his body was very tenatious: it represents a *Goldsmiths Forciples*, with the which he is wont to take a *Crucible* out of the fire, he shutting his hinder, feet as those *Forciples* are to be shut, and in that posture if you shake him, and tumble him never so much, he will rest and not move, like a dead thing ; sometimes a whole day together, but when he crept forward, he took long strides, stretching out himselfe at full length right forwards, drawing up the hinder part of his body to his fore part, like a Loop, or the Buckle of a Belt, and thus he walked.

The 25 of th. *October*, he put off a Skin, as many other Catterpillars use to doe, when they are about to change, and hung with his head downwards, from six a Clock in the morning, untill eight at night, at which time he was alwayes wont to creep abroad, he eat at nights and slept in the day time.

Now when the *Elder* leaves were fallen, and being dry, had lost their former taste and virtue, he refused to eat any more, the 19th af *November*.
he fasted all Winter, doing nothing else but shifting his
place

place every Night ; one Night hanging with his head downewards, and another with it upwards, alwayes his body at its full ftretch and length , and thus he paffed the time away. All the time I had him by me I could never perceive any fignes of life in him o'th day time, except that Day when he was put into my Hands, and thofe fmall Signes that were, I Obferved by the benifit of a *Light*.

The 20th. of *March*, the Year following, I fet before him the Buds of *Elder*, if now perchance he had a mind to eat, after fo long fafting, but he would not taft them. The fecond of *May*, I gave him the full grown leaves of *Elder*, and at length I found him eating of them.

But fo it was, that 3 Dayes before he fell to eating a-gaine, he had drawn a *Thred* along the *fand* , which I layed for him, out of the *Saliva* of his mouth, and that thred was a long one, and much fmall Sand ftuck to it, like fo many threded Beades, by the help of his Forfeet he drew this thred to his mouth againe, and fwallow-ed the fand, as I gueffed to change his fick ftomack, as certain *Eels* fwallow Prickley fmall Fifh for their ftomacks fake.

And that alfo was worth the noting in him, that he was grown a third part biggar in thicknefs and length, fince he eat, then he was, when I firft had him, that he alfo roled his body in the Sand, and tued him there, as oft as he eat.

The 16th. of *May*, he faftened the hinder parts of his body to the Glaffe, which I kept him in, his head hang-ing downewards , and he clothed himfelfe with leaves, and a Web woven by him round about his body, in the forme of a little Bag or Net : into which Bag, (changing) he put off his Skinn, and let go his hold from the Glaffe, to which, with his hinder part of his body he held faft, he let himfelf fall with his whole body, and in this pofture he betook himfelf to change.

The day after he had that forme, which is expreffed

in

in that *Table* : the 24th. of *June*, came forth an unusuall
and wonderfull *Butterfly* , of a Yellow Colour, golden
haired, which (very much shaking its Body) dyed the
5th. of *July*.

It was well our Author *lighted upon the leaves that
these* Catterpillers *would feed on : though that was done by
imagination, more then reall resemblance*

*It seemes reasonable to me, that he who would effectually
prevent the Worme eating of Ships, should well understand
the* History *of what Insect it is, that infests them; and particularly the manner of its feeding.* Wormes, *most certainly
have* Peculiar palats, *as well, as the more perfect* Animalls,
and will rather die, than eat such and such an Herb, *or*
Wood : *So that the* Wood *that by experience, they shall be
found, to refuse to eat : that is the Wood, which is the proper
materiall to* Sheath, *or* Plank *Ships with. And this being so
grand a* Desideratum, *it would be worth trying all the
numerous varietyes of* Timber Trees *which* Europe, *or
the* Indies *afford, which might be easilie done by Tallies
of them fastened to a* Buoy, *in the most dangerous Rivers,
or Bays, as I have noted above.*

Number. 11.

God.P. 2.
*T*ab.43.

The *Catterpillar* of the 11th. Table eats *Pare tree-leaves*,
but is most delighted with *Rose-tree leaves*, they are not
very greedy, and eating by intervalls they lie quiet, and
sometimes extend themselves, having fast hold of the
leafe, with the hinder part of their body. Being full, they
rest in the forme you see exprest in the Table: They
care not much to change their place, and go not from

D   leafe

leafe to leafe, as is the way of many *Catterpillars*, they willingly leave not the leafe or twig, before they have devowred all : They are elegantly coloured, the upper part of their body being darke, the under yellowifh, but not fo faire after meat, as before ; becaufe they are of a very thin fkin, and the Green food , they take, is feen through, and doth obfcure the various colours, which addorne them.

Before they betake themfelves to Change, the upper part of their body often turns redifh.

This *Catterpillar* fed by me fome time, began its change the 19th. of *June*, in that very *Forme*, which you fee him expreffed in. The 27th. of *November* came forth a leffe beautifull *Butterfly*, marked with many black fpots in the Back, like a *Marbled Book*, as the Fafhion now is; This lived a hungry life for 4 Dayes, I not knowing what to feed it with.

I had another of thefe *Catterpillars*; out of the hinder part of whofe Body, before he betook himfelf to change, 2 little worms crawled forth the 18th. of *May*, of a dufkey colour ; within 2 Hours they took the forme of Eggs, coloured like the Mother *Catterpillar*. The 9th. of *June*, the fame Yeare, came forth of thefe Eggs, two *Flyes*, which within two Dayes, fhaking and trembling, died and about that time alfo dyed the *Catterpillar* it felf. After that the *Catterpillar* had fpawned thofe worms fhe dwinled away, pined, and never eate more.

Thefe 2 Wormes, *I fay, were the difguifes of certain* Ichnenmones, *and not* Flefh-Flyes, *for the very reafon, that the* Catterpillar *furvived their eruption,* Flyes *feeding on* Carrion *or* Putrified *bodies.*

Number. 12.

G. P. 1.
Tab. 35.

The *Catterpillar*, of the 12th. Table feeds on the leaves of the *Black-curran-buſh*, it is of an other ſhape then the ordinary *Catterpillar*, which are wont to have feet in the middle of their Bodies ; but this has them at the extremities , and are very ſtrong with them, when they change place, and creep from bough to bough, they firmly lay hold of the bough with their feet, and lift up their body like a *Snake*, and ſo try for another place to remove to, they hold ſo faſt by their feet, that you can ſcarſe pluck them from the thing, they adhear to.

This *Catterpillar* changed (having firſt exactly purged its Body) the firſt of *Aprill*, and the 24th. of *July*, came forth the *Butterfly*, expreſſed in the 12th. Table : tender, and a ſhort lived *Animal*.

Number. 13.

G. P. 3.
Tab. 5. K.

This green *Catterpillar*, of the 13th. Table, eats the Leaves of the *Elder-Tree*. When its full, it extends its Body forcibly the whole length, and in that poſture reſts untill the next day. As it creeps, it joynes its hinder parts to its fore-parts, like a *Loop*, and then extends it ſelfe forwards, and then gathers it ſelfe up againe, and ſo on to its journeys end.

It changed the 6th. of *October*, 1663. And the firſt of *May*, the Year following. 1664. It appeared in the ſhape of a moſt beautifull *Butterfly* expreſſed in the 13th. Table, I choſe the moſt Elegant of them to deſigne that by,

D 2   becauſe becaufe there is fome difference in fuch, as are bread of this *Catterpillar*. This *Butterfly* was very nimble, and of a fwift flight.

Number. 14.

G.P. 2.
Tab. 21.

The *Catterpillar* of the 14th. Table, expreft to the life, are great Lovers of green Garden *Lettice*, but will not (like many other *Catterpillars*) touch the Plant withered.

This *Catterpillar* began its change the 8th. of *June*, in the *Forme* expreffed.

About the beginning of *Auguft*, came forth a *Butterfly*, hoary colour, having a long Stinge in his fore-head, or an extended *Ray*, which lived but one day fafting, I knew not what to feed him with.

But I have Obferved this kind of *Butterfly* to fly about Flowers, dipping that long Sting, or Tongue into the flowers, and fo to fuck out the Honey, or fweet juice, not fetling or lighting upon the Flower, but doing of it as they fly; for if they fetled, their Legs were two fhort to fuffer them to ufe their *Tongue*; and this I take to be the caufe, of their conftant taking their nourifhment in flight, and not alighting, or fetling upon the flower,

There is a fort of *Gnatts*, bread in Trees, which in like manner are armed with long Stings, and can ufe them fiting, becaufe they have long Feet, efpecially in the hind-moft feet, fo that he can fasten his Sting where, and how he pleafes ; but the naturall fhort feet of thefe *Butterflyes*, hinder their feeding, fave when they are upon the Wing.

This

This Butterfly *is an exceeding swift flyer, he is very frequent in* July, *in Gardens amongst* Gilly-flowers *and* Pinks. *I didonce take a small* Species, *of them exceeding beautifull; in* Morgans Garden *in* London: *I us't to call them the* Hawk-Flye, *for their exceeding swiftnesse.*

Our Author, *in my opinion has well assigned the reason of their Feeding Flying, to the disproportion betwixt their Legs and Tongue, or Pipe.*

Number. 15.

The *Catterpillar* of the 15ᵗʰ. *Table*, runs swiftly ; for every stride is the whole length of his body, going like a paire of Compasses. It is found for the most part about the *Thlaspi* kind of Plants , which kind of Herbs it greedily devowers : It feeds by Night only, and moves not by day ; and although you do disturb and molest it, yet it will keep it selfe quiet without any manifest motion.

It changed the 11ᵗʰ. of *August*, 1664. And the first of *September* it appeared again, in the shape of the *Buutterfly*, expressed in the 7ᵗʰ. *Table*, of a rapid flight , which yet it wou'd not be brought to, but by the light of a Candel, or some such Artificall Light.

In Flying it vsed a strange motion of its Body , as though it was perpetually circumvolved, with its Head downwards, and in a Circle.

It

It Feeds by Night only :] *These* Animalls, *as well as others, have a naturall instinct, to preserve themselves, and feed only, when they are most secure, and the* Insectivorous *Birds at Roust.*

Number. 16.

The *Catterpillar* of the 16th. *Table*, eateth *Hyssop*, and is found chiefly about that Plant, when in flower, but if you touch the Plant, it falls to the ground, and thus hides it selfe.

After I had a long time fed this *Catterpillar*, with *Hyssop*, it changed the 7th. of *August*, and when it had been some dayes in that condition, Three Wormes broke out through the skin of it; which in a very short moment were changed into Three Eggs. And out of every one of these Eggs, the 8th. day of *September*, came forth a Fly, as I have expressed them exactly in the 16th. *Table*, the Fly lived not above three dayes.

These were Flesh-flyes, *and consequently the* Chrysalis *was Rotten, and this* History *not finished ; because we have not the* Butterflyes, *which in reason we ought to expect, the disguise declaring as much to us. I am at a losse*
how

9.

12.

15.

16.

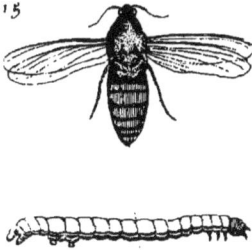

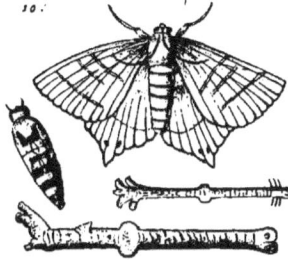

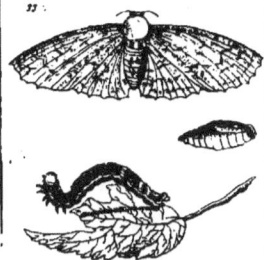

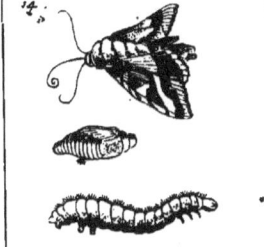

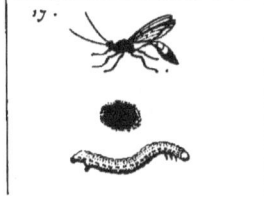

(23)

how it comes to paſſ, that in our Northen *parts of* Europe, *where ſuch Plants as* Hyſſop, *&c. Doe not naturally grow, there ſhould be found* Inſects *to feed on them; we muſt either ſay, that they came into our parts by degrees, that is the Plant being cultivated at firſt, not far off the place, where its Native Soile is, the* Butterfly *ſtrayed, and found it out there, and ſo on; or, (which is moſt probable) this* Catterpillar *will feed of other Plants, as well as* Hyſſop, *and ſo the* Catterpillar *is no Stranger to the Soyle, though the Plant be; This I ſay is more probable, then that the* Butterfly *and Plant ſhou'd both be Strangers, for I cannot think, that this or any* Animall *elſe is* Spontaneouſly *produced by the Plant, or any cauſe elſe whatſoever, but the* Animall Parant.

G. P. I.
Tab. 71.

Number. 17.

The *Catterpillar* of the 17ᵗʰ. *Table*, is all green, like the *Curran-tree* leaves it feeds on: it Eats thoſe leaves, begining in the middle, where it greedily feeds, but touches not the outſides of the leaves, when full, they lift themſelves up, and then extending themſelves they reſt; continuing in this poſture three Hours, untill the meat be digeſted, then they excerne, what has been put over, and fall too againe; they never drink as many other *Catterpillars* do, that I cou'd obſerve, therefore muſt you feed them with freſh leaves conſtantly, if you wou'd know what they will come to. It purgeth its ſelfe from all its excrements (as is the cuſtom of *Catterpillars*) and began its change, gathering ſome leaves I had given it, Glewing them together by a Slimy humour, (from its mouth) about its body.

This (*Catterpillar*) expreſſed in the *Table*, remained
ed

ed without motion, from the 4.th of *June*. till the 5th of *May*, the Year following, that is a 11 Months and a Day: at which time came forth a *Fly*, with a Black body, all but the upper part of the Taile *Yellow*. This as soon as it came to Light, rested a while, untill the aire had dryed and hardened it, and then it pruned its Wings and flicked its Body over, and prepared its felfe for Flight,

Our Country-people call thefe kinds of *Catterpillars*, *Surveyours* (*Geometrε*) becaufe of their Gate, which is like a Pole turned over and over, as when one meafurs Land.

This Fly, *is one fingle pretty large* Ichneumon, *or* Slen-der Waspe, *and therefore this* Hiftory *, as well as the precedent, are imperfeɔ; becaufe thefe are but* By Births *and not the* Butterflyes, *which ought to be expeɔed, and of which thefe* Catterpillars, *and their* Chryfalis *are but the difguifes.*

Section.

Section. III.

Of *Butterflyes* fitting with *Hanging Wings*, and clapt to their Bodies, like the Wings of *Birds*.

This 3*d. Section* of *Butterflyes* is very large and contaÍnes the *Hiſtory* of a numerous *Tribe*: I ſhould have had great ſatiſfaction, if I could have taken the *Subordinate Genuſes*, from any differences remarkable in the *Butterflyes* themſelves, and not have been forc'd to order them by certaine common Notes Obſervable in the *Diſguiſes* of the *Catterpillars*: But the want of Particular diſcriptions, which our *Author* has wholy omitted, has neceſſiated me to this method; from the bare *Figures*, little of certainty can be gather'd. We indeed Obſerve the *Antennæ*, of ſome *Butterflyes* to be ſmall like *Wiers*, others *Plumous*: Again ſome to be very full of *Feathers* upon their *ſhoulders*, others upon their *Heads* &c. But theſe differances are neither certain nor common enough: Theſe of the *Catterpillar* on the contrary, are more *notable*, ſuch as their be *Horned*, *Hairy*, *Naked*, *Great*, *Little*, &c. At preſent we muſt be content with theſe, wich is moſt pardonable, then the putting the *Catterpiller* in one *Chapter* (as *Mufet* has done) and the *Butterflyes* in another, as though they had little relation to one another.

Number. 18.

God, P. 3.
Tab. 1.

The *Catterpillar* of the 18th. *Table*, is to be found in *Elder Groves*, ſo long as the leaves are tender, and full of juyce, but when they begin to be hard and Withered

E   by by the heat of the *Sun* , it leaves them and frequents other places, it feemes to be peaceable and Tame, for it refifts not, but flyes if it be injured.

It began to Change the 6th. of *Oæober,* and the 25th. of *Aprill ,* the Year following 1664. It appeared in the fhape of a Blew *Catterpillar,* mixt with a Redifh brown, in the *Hinder-parts,* of his Body.

It lived from the 25th. of *Aprill ,* untill the 2d. of *May,* without moving its Body from the place, and fo it died.

But before it died, it layd certain *Eggs,* of a notable Green colour, which our Peope call *Spanifh-Green.*

The Eggs of Butterflyes, are to be found in their bodies. whileft they are in the Difguife of a Chryfalis, and therefore are undoubtedly effentiall parts of the Female, as much as her Legs and Wings, and in no wife generated by the Male. As foone as they have thrown off all manner of Difguife, they are ready for Copulation with the Male ; but if they chance to be Unmafked alone; and not in the company of a Male, they lay inftantly their Eggs without Copulation; though probably fuch Eggs are not Feconde and will not come to hatch.

G. P. 3.
Tab. L.

Number. 19.

The *Catterpillar* of the 19th. *Table* eates the Succulent and frefh gathered leaves of *Oziers,* thefe leaves that are white and long. It is a flow goer.

He changed the 10th. of *Oæober,* 1662. Aud the firft of *May,* of the Year following 1694. A very great *Butterfly* came forth. This *Butterfly,* toucht with a *Pin,* fell down *Precipitately,* like one dead, and without motion, fo that I thought him dead, of which that I might be certain, I ftuck him through with a *Pin,* and yet he fhewed no fignes of life, but by and by he Expanded his Wings and endeavoured to Fly away, but he had not ftrength, he livid untill the 11th. of *May,* 1664.   **Num-**

Number. 20.

The *Catterpillar* of the 20th. *Table*, feeds flowly of the leaves of the *Willow*, and is flow of gate, it abftained from Meat the 24th. of *Auguft*, and began to *Weave* certain Threads, but defifted, and often changed its place, untill the 14th. of *Oftober*, at which time it changed its Green Colour, into a brown one and dyed, I think that this *Catterpillar* abideth in hollow *Willow Trees*, and is turned into a *Butterfly* in *Spring* : but guefles are uncertain, and as yet I could make noe experiment.

Again of the fame. N. 20. B.

The *Catterpillar* exprefled in *Number* 20th. Eats Night and Day of the whiteft leaves of the *Willow*, and fupper over, he withdraws, and hides his Head like a *Tortoife* within his body, leaft the night Aire or Raine fhould hurt him, after that I had obferved him to abftaine from his ordinary food, I put him into a Glafle halfe filled with Earth, throwing in fome Fragments of *Willow* leaves, thefe leaves he forthwith crumbled to duft, and fo kneaded them with the Slimy juice of his Body, that he made him a very convenient place for the Change, the Houfe was like *Willow-wood*, but harder far, fo that it could fcarfe be pierced with a Knife, and one would hardly guefle a *Catterpillar* was within it.

E 2   Before

Before the change, in the space of two hours, he lost the beauty of his Colour, and was then of a Liver Colour brownish: As though he was to dye.

He remained in this Change from the 20th. of *August*, 1663. untill the 11th. of *Aprill* 1664. And then came forth of that most harde Shell, a *Butterfly* of an *Ash* Colour diluted, whilst he lived, he scarse moved, and yet run thorough with the Needle, he lived untill the 24th. of *May*.

G. P. 2.
Tab. 37. *Again of the same* N. 20. c.

This is a Rare and unusuall *Catterpillar*, he is for the most part to be found on the Sand Hills, which lie along the Coasts of *Holland* and *Zealand*, and feeds on the leaves of a kind of *Palme* (called by our People *Duin Palme*,) (perhaps *Myrtus Brabantica*,) this Herb was very gratefull to him, but I that dwelt farr off the Sands, could hardly get it, wherefore I taught him to feed of certain *Osiers* or *Willows*, which seemed to be soma thing like the Plant, in Colour ; and drynesse.

This *Catterpillar* had two Tayles and when he was Vex.d, out of each Taile he put forth a *Red Sting*, which he bent and brandished, trembling as though angury, and again drew them in, and also he could cunningly draw up his Head, and hide it like *Tortoises*.

He changed the 4th. of *September*, under certain *Willow* leaves, which he had knit close together by certain Threds of his own Spinning in the former of five little Cells, as you see in it expressed in the *Table*, which when It uc ed with a Knife, it felt as harde, as a Stone. In this manner he lay and moved not, or shewed any signes of Life in Nine Months, and Fifteen Dayes, for the 20th.

of

of *June*, the Year following, out of the five Cells came forth five *Flyes:* which *Cells* when I had opned I found nothing at all in them, not so much as a Skin, or the Taile or Feet, I wondered all was consumed, I kept these Flyes for some time alive, and when I expressed one of them in the *Table*, I gave them their libertie, not thinking it necessary to Paint the rest.

These three Tables *of our* Authors, *I have made one, and put the* Histories *together: because in my judgment they are about one* Species *of* Butterfly.

(2) *We must commend the industry of our* Author *in this place, because not having succeeded, when he first kept the* Catterpillar, *he (the next time he lighted upon it) supplyed it with Earth and fragments of Leaves, in order to its more convenient change, and it must be observed, that many* Catterpillers *require a* Forrian *matter to make a* Foliculus *of; for the security of the* Chrysalis, *as Leaves, Earth, &c. Though indeed, others Spinn themselves one, out of their owne Bowels; or use their Haire for that purpose.*

(3) *The* History *is compleate in the second* History, *where are the two Disguises of the* Butterfly *it selfe.*

The 3ᵈ· History *presents us with a* By-birth, *but indeed a very Rare and curious one: here we note that these five* Cells *were the Workman-ship of the five* Ichneumon Wormes, *and very probably they made use of the* Catter-pillar *for stuffe to make them* Foliculi *out off.: Just as he sayes in the* 2ᵈ· History, *the* Catterpillar *crumbled the Fragments of Leaves and Earth, and Kneaded them into a* Foliculus, *of a strange hardness. He hath not Figured that* Foliculus *in the* 2ᵈ· Table, *but took the* Chrysalis *out of it, and Painted that alone.*

E 3·   We

We now come to treat of, or order all the *not Harie* and *Tuberous Catterpillars*, of our *Author*, and such as have *Hooks*, and Hornes, or *Bunches* any where about them.

Number. 21.

The *Catterpillar* which you see expressed in the 21th. *Table*, feeds on *Alder leaves* ; Upon its back stands up two yellow Hooks, the rest also of his body is eged of a pale Yellow ; but before he changed, he put on an abscure Green colour, and the two Hooks fell within his body, so that now he was every where plain and smooth.

He began to change the 6th. of *October*, 1663. And appeared in the forme of a *Butterfly* the 27th. of *Aprill*, the year following, which for the great diverfity of colours, I could scarse describe, it was most lively, and flew the nights through almost without ceasing, it lived untill the 5th. of *May*, when it died with extended wings.

These sorts of Catterpillars, *with Hooks on the middle of their bodies, I do not know that I have yet met with, and cannot therefore guesse at the use of them, if it be not for defence only, as* the Author *expresses it in the next* History. Number.

Number. 22.

G.P. 2.
Tab. 38.

This *Catterpillar* of the 22d. *Table*, was found in *Flan-ders* at *de Groeds*, a Village over againſt *Fluſhing*, upon a *Willow bough*, in that very poſture it is expreſſed in the *Table*, it feeds of *Willow-leaves*, but eats very ſloly, and after eating compoſeth it ſelfe again, in the poſture you ſee it *Figured*, in the hinder part of its Body repreſents the Head and Beard of a , *Goat*, and it doubls or draws back its foreparts, of or to the hinder parts; when you touch it, it ſtrikes at you with Head and Taile, as though it was angry: In the Back it has two Hooks, with which it ſtoutly defends it ſelfe, ſo that it ſcares all Creatures, that ſees it: when it eats, you would ſay its head was bound to its body, by a ſlender thred, not unlike the body of a *Spider*.

The 1ſt. Of *September*, it reſted and began to change, in the forme expreſſed in the *Table*, and after 22 Dayes, came forth a *Butterfly*, diſtinguiſht with variety of co-lours, Elegantly.

This *Butterfly*, before its death, layed its Eggs of a Green colour, in divers figures and faſhons expreſſed in the *Table*, otherwiſe then the reſt of *Catterpillars* doe.

It lived with me from the 21ſt. of *September*, untill the 3d. of *October*.

This.

This Butterfly, *layd its Eggs in certain broken Linkes,*
or Chaines, *as is expreſt in the* Table ; *But I ſuppoſe ſhe*
would have layed them in another manner, if the Male had
firſt made them Proliffick , *however I affirme nothing :*
having obſerved certain Butterflyes *Eggs, wrapt and*
wound in Spirall Lines, *about a* Twig ; *and moreover the*
Eggs *of* Froggs, *and of ſome* Fiſh, *are Spawned in* Chaines.

G. P. 1.
Tab. 33.

Number. 23.

The *Catterpillar* of this 23*d.* *Table*, feeds upon the
leaves of the *Sallow*, it is rare to be met with, and I
nevercou'd find any, but that one only : it daily required
freſh food.

After that I had fed it a whole Month, the knit certain
leaves together like a *Net*, in which he hid himſelfe.

And then he changed, as I have expreſſed it in the *Table*.

After he had had his fill of leaves, (as I said) he changed the 19*th.* of *July*, and the 3*d.* of *Auguſt* came forth
a *Butterfly* painted alſo in the *Table* : which lived ſix
Dayes without food.

Number

18.

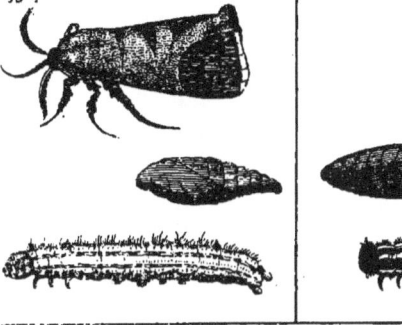

20. b

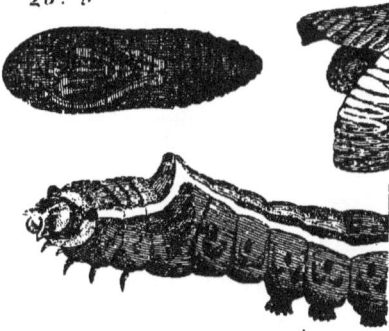

21.

2

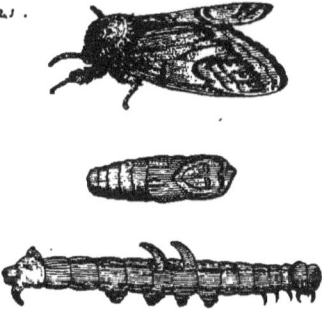

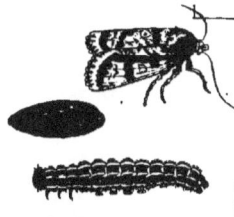
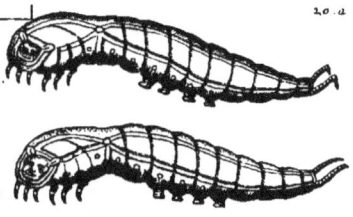

20.a

20.b

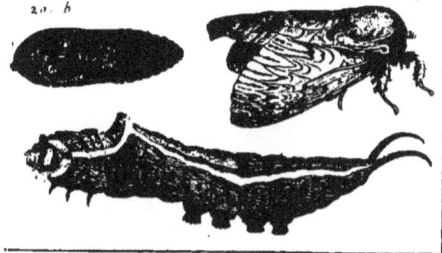

20.c

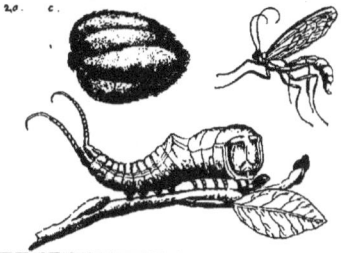

21.

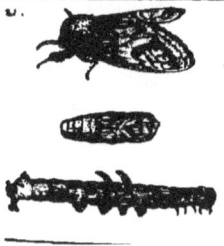

22.

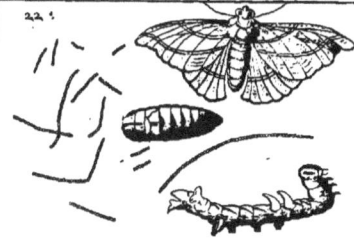

23.

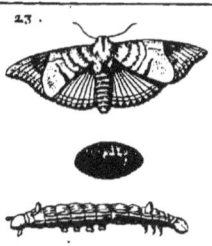

The Green *Catterpillar*, of the 24*th. Table*, feeds on *Sallow* Leaves, untill the end of Snmmer. It begins to Eat at break of Day, and eats with great greedinefs 5. or 6. Hours without ceafing, or other imployment : It never voids its excrements, unlefs when new food drives out the old. It cleaves fo faft to the *sallow leaves*, that it can fcarce be pluckt from them; but will fuffer it felfe firft to be pluckt in pieces. In the hindmoft part of its Body it has a fharp *Pinn* or *Sting* , and *Venemous*, which it vfes (as foon as it is troubled) to defend it felfe with, with a fwift agitation, and every where; that it may Wound them that hurts it, with the motion of the Sting.

The Male *Catterpillars* are Green (as we faid) and want thofe round fpots which look like Eves; the Females are not fo Beautifull, but of a more Grayifh colour.

Before that thefe *Catterpillars* change, thev move long and much, and exactly clean themfelves of all Excrements: they make themfelves no covering (as all other *Catterpillars* are diligently wont to doe) but puting off their Hackle, they are changed into *Chryfalis's* as is expreffed in the *Table*.

This change happned the 19*th.* of *September*, 1663. and the 5*th.* of *May*, the Year following 1664. came forth a lufty *Butterfly*, which lived till the 14*th.* of *May* fafting.

I could never Obferve from thefe kind of *Catterpillars* a perfect and compleat *Butterfly*, but with contracted, and as it were fcorched Wings, not to be expanded, or fit for flight. They are wont to lay their Eggs before they dye.   F   (1)

(1.) *I note particularly the observation of our* Author *of* New food driving out the Old, *which must be either that the intestines are not so very long, and many as in* Sanguineous Animalls ; *or this is done upon the account of a* New fermentation.

(2:) *Its a curious* Note, *the distinction of the Male, and Female* Catterpillars: *this would be certainly known in order to the compleating their* Historys, *for as* Malpigius *observes, the Eggs are of one colour, when layd after* Copplation *and of another without it.*

(3ly.) *A probable reason why our* Author, *never could have a compleat* Butterfly, *of this* Kind, *was, because he did not furnish the* Catterpillar, *with convenient matter for a* Foliculus, *and so in some measure starved it.*

G.P. 1.
Tab. 24.

Number. 25.

The *Catterpillar* of the 25th. *Table*, feeds of *Sallow leaves*, it is armed both before and behind, in the forehead it has a kind of hard Shield , but no Eyes, that I could Observe, and in the hindermost part of the body, it has a Prick or sting, hard and stiffe , it is of a Green colour mixed with Blew, I take it to be Venomous: if it be touhced, it moves that part of the body, which has the Sting, and does as though it did defend it selfe with it.

It cleansed it selfe, and began to change the First of *Aprill*, in the form expressed in the *Table*, in which condition

dition it remained for 55 Dayes, not obscurely, representing the Face of a Man with a Beard, and in the other parts a Child in Swadling Cloathes.

The 25*th.* of *May*, broke forth a *Butterfly*, figured in the *Table*.

This Catterpillar, *feeds also of* Liguſtrum *or* Primp leaves, *and is to be found about* Primp hedges, *with us in* England : *It is a moſt beautifull Creature.*

Number. 26.

G. P. 3.
T*ab.* Y.

I v'ſt to call the *Catterpillar* of the 26*th. Table* the *Elaphant* becauſe of its Snout, it feeds of *Vine leaves,* when it has done feeding it draws up its Head within its Body, and hides it ſelfe like a *Tortoiſe,* its very fearefull, for being touched never ſo lightly, it trembles and is aſtoniſhed, enough to cauſe the ſame affections in the beholders. It can turn its Snout very dexterouſly every way, and what its layes hold of with it, its hardly to be got from it, for it hath great ſtrength : Its very quick of hearing, for at the leaſt noiſe it feares and contracts it ſelfe, as I have often tryed.

The time of its change approaching, it abſtaines from meat, and ſeeks a fit place to reſt in, and to that purpoſe I put it ſome bits of Wood and *Vine leaves,* of which and the juice of its Body mixt, it made it ſelfe a Houſe, and changed in it, the 4*th.* of *September,* 1665. And lived in it untill the 16*th.* of *Aprill* and longer, which thing I obſerved by mine Eye and Touch, being uncertain what wou'd become of it.

When

(36)

When I touched it, and put it in the hollow of my Hand, I observed it as cold as Ice, and yet moved most strongly.

This is most certain (this Animall) did live without food from the 4th. of *September*, 1665. untill the 16th. of *Aprill*, 1668. To which its coldnesse and continuall rest, did seem to conduce very much, from which and the Tenaciousnes of its juice, it would be long in wasting, which thing is also frequently experienced in other Creatures, which rest in winter, and eat not till the Spring or summer.

This History *is imperfect, there being only the* Disguises *and not the* Butterfly : *I am apt to believe the* Chrysalis *dyed, by being taken out of its* Foliculus, *and therefore in vain the* Author *expected to see it hatched, or a* Butterfly *breakforth.*

As to its coldnesse ; I can witnesse that in the depth of Winter, *and in the very deep* Snow, *I have found both* Catterpillars *and* Hexapode Worms, *lying upon the* Snow, *and therefore have crawled out upon it. I say these* Catterpillars *were so hard Frozen that thrown against a* Glasse , *they made a sound like stones ; but put under the* Glasse *and set before the* Fire, *did quickly crawl about, and bestir themselves nimbly to get away.*

O. P. 3.
Tab. V.

Number. 27.

This kind of *Catterpillar* expressed in the 27th *Table*: Is amongst *Corne*, and there feeds, of a certain Bind-weed, (*Hedera minor*, called in *Dutch*, *Wee windé*) and

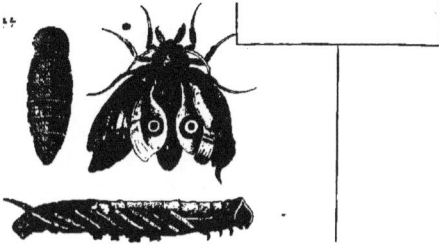

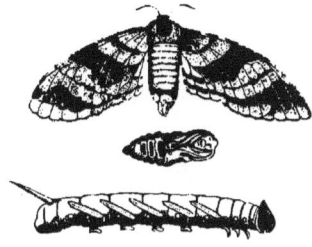

and is at leaſt two Hours at one meale: Its excrements are
of a determinate figure, expreſſed in the *Table*, it is of
divers colours, with two black Lines on its back, betwixt which two, is there one of Green, the reſt of its
Body is Greeniſh, in the hinder part of its Body is there a
Red and crooked ſharp Horne, and in ſome places is there
a mixture of White and Blew, it is of a quick Eare and
touch, for upon the leaſt noiſe, it will not move of a long
time. The *Author* prevented by Death left no more.

This Hiſtory *is confeſſedly imperfect, there being nothing but the* Diſguiſes *of a* Butterfly *expreſſed. I cou'd
have wiſhed the* Author *had taken all along the pains
he has here ſhewed, in more carefully deſcribing in
words the* Painting *of the* Catterpillar, *but he relyed upon
his excellent pencill too much, for he had elſe leiſure and
opportunity enough to have done it. This* Catterpillar *is*
Cornigerouſe, *and therefore we put him amongſt his like.*

2. *As for the determinate Figure of the* Excrements *of
this* Catterpillar; *if the* Author *had well obſerved it, he
might have found that particular common to many* Species
of Catterpillars. *I take the cauſe of it to be, the Figure of ſome
parts of the* Colon, *or* Gut *wherein theſe excrements are
baked and molded.*

Number. 28.

G. P. 3,
Tab. Z.

The *Butterfly,* of the 28th. *Table :* was ſent from *Paris*
to the *Author,* for its beauty and vaſtneſſe, in which it
far exceeds all others, to the end upon occaſion, he might

find.

find out his Origin. It was taken in the Kings Garden, and given to Dr. *Borellius*, the *States Embassador* : Who carefully transmitted it to him.

In my Judgment this Butterfly *will be found to have a Disguise, like some of the next precedent : And to have been changed from one of the* Cornigerous Catterpillars : *but it is improper to talke here of* Origin, *because it is perfect of it selfe, and is the* Mother *of the* Catterpillar, *which it is changed from.*

We Proceed in the *Histories*, of the smooth and not Hairy *Catterpillars*.

G. P. 3.
Tab. F.

Number. 29.

The *Catterpillar* of the 29*th. Table*, feeds of *Colly-Flowers*. One in a corner hid under a Paper was found dead, out of whose Body the 10*th.* of *September*, 1663. 12. *Wormes* came forth, which in the space of about four Hours were turned into so many *Eggs*, the 2*d.* of *July*,
the

the Year following 1664. out thefe 12 *Eggs*, came 12 *Gray Flyes* [one of them expreffed in the *Table*.]

Another died the 12*th* of *September*, whofe relicks were changed into a Worme the 17*th*. of the fame Month, which Worme broke forth of the fkin of the faid *Catterpillar*, with great ftrength and force.

A 3*d*. *Catterpillar* of the fame *Species*, changed the 25*th*. of *September*, 1663. and the 8*th*. of *May*, the Yeare following, it appeared in the forme of a *Butterfly*, of *a* flow gate, and no pleafant colour, it lived fafting Nine Dayes.

Note the Firft Catterpillar, *our* Author *plainly fays*, Died, *and the Carcafe. undoubtedly was* Fly-blown, *fo that the* 12 *Wormes were Maggots, of fome fmall* Flefh-flyes, *and accordingly they proved, one of them being Figured in the* Table.

Againe we believe the fame of the Worme, which broke out of the dead Body of the other, but the Author *being filent, what became of it, we fhall not concerne our felves to gueffe of what Race it was.*

Number. 30.

G. P. I.
Tab. 67.

The *Catterpillar* of the 30*th*. *Table*, feeds of divers Herbs, amongft the reft; it chiefly delighted in *Ground Juy*, but it is delicate and choice in its meat, for it will eat none but frefh gathered : it changed its Skin oft times, for it feemed fo hard, that the Body being notably en-
creafed

creafed in bulk, it was neceffary that the fkin fhou'd be broken; and another grew under it.

Its fkin being caft, it did not move for one whole day. nor cou'd it feed, becaufe of the tendernefs of the new fkin which was hardened by the Aire.

It changed the 5th. of *Auguft*, as is expreft in the *Table*, and the 26 of the fame Month came forth a *Butterflys* figured in the *Table*.

Thefe *Butterflye*, are wont to fly about chiefly in the night, but are careful to avoid the flame of Fire and Candle which thing happens otherwife to the reft of *Butterflyes*.

This *Butterfly*, Lived 6 Dayes without food.

Note here our Author gives a good reafon of the cafting of the Skins of Catterpillers, faying that it grew hard and was not further to be extended, fo that of neceffity the growth of the Animall cracks it : The like I have Obferved in Spiders : This alfo holds good in the Hornes of Animalls, which cafte them, as Staggs : Which when they are at their full growth are deprived of all further nourifhment, become hard, and are fuceeded by others which pullulate ; the like is obferved in the Hair of Animalls, and the leaves of Plants.

It is no leffe certain that all the Skins are one under the other, each fucceffively Pullulating, as in the leaves of Plants.

G. P. 3.
Tab. 25.

Number. 31.

I have often obferved the feed of *Heliotropium* fhed upon the ground, to be, preferved all Winter, though never fo Colde and to *Germinate* in the Spring : The *Catterpillar*,

terpillar, of the 31ft. Table, eats this Herbe, it lyes under ground ith Day, and comes not out untill the Evening to feek food, I kept this Catterpillar under a Glaffe, but I experienced that it never moved, or eate any thinge fave ith evening.

After that it had duly cleanfed, it felfe, it began its change the 17th. of Auguft, & remained in that condition, untill the 9th. of June, in the Year following, fo that it was like a dead thing without food almoft 10 Munths.

Afterwards came forth a Butterfly, marked on each Winge, with the Letter o. And Figured in the Table.

Here we muft note, how neceffary it is, in order to the compleating of Naturall Hiftory, that our Naturallift fhou'd be well skilled in Plants: Viz. The Food of moft Infects. Heliotrupium, is a Name given to many Plants, as to one Species of Tithemall, alfo to the great Indian Marigolde; again to other Plants, Famous for Dying a Blew Colour, &c. So that the Author has left us in the Darke for the Food of this Infect; for want of a more particular Title of this Plant.

Number. 32.

G. P. 1.
Tab. 42.

As foon as Mulbury-leaves are fpred, the Catterpillar (called Bombyx) of the 32 Table, is Hatched, and feeds on them, although in their Infancy they will Eat Wilde

G   Cichorie

Cichory and *Lettice*, for the *Mulbury* is late in putting forth its leaves.

Before that thefe *Catterpillars* makes Silke, they cleanfe themfelves from all Excrements.

One Hundred and Threefcore *Catterpillars* (160.) are wont to make one Ounce and 80. Graines of Silke.

The *Catterpillar*, Figured in the *Table*, changed the 14th. of *September*, and the 24th. of the fame Month, that is in the fpace of 10 Dayes, it broke forth a white *Butterfly*.

Before thefe *Butterflys* Couple, both Male and Female are exactly purged, and then they are exactly Coupled by the Tayles end ; after that the Female layes a 160. Egges which are all Infæcund, and wither to nothing unlefs the Female couple with the Male.

Sometimes they do againe Couple after Laying, but they Lay no Eggs, for the *Butterflys* are wont to Dye within Fourteen Dayes.

See M A L P I G I V S D E B O M B Y C E.

G. P. 2.
Tab. 20.

Number. 33.

The *Catterpillar* of the 33d. *Table*, Eats *Plumtree leaves* it's a little eater, and fleeps much, you will often finde Two of them together at reft with their Heads inward bent to the middle of their Bodies.

Both of them two companions began to change the 25th. of *May*, as is expreffed in the *Table*, and from each of them came forth a *Butterfly*, after 31 Dayes, that is the 26th. of *June*, of a Yelloifh colour, and which wou'd not fly in the Day time.   Thefe

These kind of *Butterflyes*, feem to be weak fighted; they care not for flying, but runn fwiftly, and they runn into fome darke corner, not caring for the light.

They live of Honey gathered out of Flowers, and therfore have a long Tongue to fetch it out of the Bottoms of flowers.

This Tongue they carefully, when they runn, roll up, to preferve it from injuries, they lye in darke places in Winter time.

Number. 34.

O. P. 2. T.ab. 24.

The *Catterpillar* of the 34*th. Table*, Eats of the leaves of that *Worm wood* chiefly, which grows upon the Sea banks, I found it fitting upon that Plant, in an *Ifland* of *Flanders*, near *Axella*, and *Neoza*, they are their whole Body throughout, exactly of the colour of Sea *Wormwood*, they are fierce, and Subtle and ftrike at whatever touches them, they hold not very faft, they runn fwiftly, where they feed they fcatter much *Worm-wood leaves*, and when they are neverfo lightly touched (of which they are exceeding impatient, tremble and feare) they caft themfelves down alfo, and hide themfelves there, where you can fcarce difcern them; becaufe they are moft exactly coloured like the Plant they feed on : they fave their Head by covering of it with the hinder part of their Body : in the Face they are tender (as I have experienced) and foon killed with a fmall hurt in that part, and therefore are very carefull of it.

Again I do wonder at their fo well enduring of colde

and

and wet that you cant carce kill them, and although they seem to be quite stormed, yet they will revive, for I cast some of these *Catterpillars* into cold water, and Kept them 12 Hours in it, then I took them forth stiffe with colde and extended, so that I cou'd not discerne the least signes of Life or motion in them; but exposing them to the Sun, within halfe an houre they came to themselves and cou'd not be discerned from their fellows that had not beeu steeped. If you in like manner Drown Flyes, in Bear or Water, and let them lye in it all Night, and having taken them out, sprinkle them well over with the Powder of Chalke, they will by and by creep away, which seemed Dead, the heate of the Chalke which I look upon as a kind of unquenched Lime, put Life into the Flyes, as the heat of the Sun into the *Catterpillars*, the Sun causing the Vitall juice to returne into, and passe the benummed Members.

Further it is to be observed, that this steeping in cold Water, gaine these *Catterpillars* a good stomack, and made them eat more greedily then their fellows which were not served so, as men seem to have better stomacks in Winter, then in the heat of Summer.

These *Catterpillars*, make themselves commodious nests to change in, of *Worm-wood* scattered upon the ground, and shutting up themselves, they change therein.

This *Catterpillar* changed the 4th. of *September*, and lay in that condition (figured in the *Table*) 10 Months, 18 Days.

And the 19th. of *July*, came forth a *Butterfly*, of a wonderfull shape, and accouterment : it lived not above foure Dayes, for I knew not what to feed it with, that it wou'd Eate.

(45)

Number. 35.

G.P.2.
Tab.27.

The *Catterpillar*, of 35*th. Table*, delights to feed of *Brambles*, and of the *Vine*, he is flow of motion and gate, and yet will feeme difpleafed if yov injure him, and defend himfelfe againft violence; I fed this the fpace of foure Months, which time being over, he made himfelfe a little Houfe of fair *Paper*, the leaves of *Balme* and *Spittle* of his owne, to change more fafly in.

He changed the 12*th.* of *September*: and remained in that condition *Figured* in this *Table*, Eight Months compleat no more nor leffe: for the 12*th.* of *May*, the Year following came forth a *Butterfly*, not unbeautifull, but impatient of Hunger, for I knew not what to Feed it with, and Dyed within one Day.

Number. 36.

G. P.2.
Tab. 28.

The *Catterpillar* of the 36*th. Table*, is not given to any peculiar food, as moft other *Catterpillars* are, but Eats of every Herbe you give it, it greedily Eats *Rofe leaves*, *Mints*, wilde *Awrache*, and many other Herbs; he never ftired it h Day time, but as oft as I obferved him at night, I found him Eating, and he feldom left any thing of that I gave him: He retired when, he had done Eating, under a Dry leafe, rather then a Green one.

G 3

This

This *Catterpillar* when he had spent a whole Day in Creepeing about, he rested under certain leaves ; and there changed the 23*d.* of *September*, and remained in that condition, Figured in the *Table*, 9 Months and 4 Days ; for the last of *June*, the year following, came forth a most beautifull *Butterfly*, painted with divers colours wonderfully.

There are some of these *Catterpillars*, twice as big and as long as others, I took a lesser *Catterpillar* of this sorte, and I fed it long, but as the season of the Year declined, and leaves grew harde, he could not Eat them, he not withstanding had endeavoured to change, but nothing came of him, and I left my experiment.

But standing at my Doors, set about with *Elmes*, I saw a great company of *Catterpillars* upon an *Elme* bough upon which withont doubt their Mothers had laid them.

These little *Catterpillers* let themselves down, by certain single Threds out of their mouthes, they were all exceeding Green, and Pelluced, I took them and fed them with *Elme leaves* under a Glass, they all Housed themselves with houses of Sand and Spittle, against the cold and of winter.   In Spring I let them seek their own food.

I have not all the satisfaction I cou'd wish, in Translating *this* History, *I find it so imperfect and disordered: undoubtedly the* Latin Translator *is much to blame, I have mended the* Text, *where it speakes of the great* Catterpillars *being the* Mother *of the little ones; then which nothing can be more absurd: again we have here an* Aurelia, *put for a* Butterfly *&c. If it were matter of Fact and positively asserted by our* Author, *as his Observation, I shou'd not have alter it, but its expressed, as a conceit only.*

Number.

Number. 37.

The *Catterpillars* of the 37*th. Table* Are rare : I found this in a *Sand-hill*, the 23*d.* of *June*, sitting upon Grasse, which grows there (called in *Dutch Duin grass*) and I took it home with me to try what cou'd become of it.

The Day after a Little *Animall* like a *Beetle*, crept forth of the hinder parts of its body, *Figured* in *Table*, 37.

The *Catterpillar* after this refused all food, and contracting and winding its body, it changed at length.

The little *Animall* lived, but stirred not much, only shaked it selfe, it changed *Snake* like, and the skin being cast, it was of a Golden colour on the fore part of its Body, and russet the hinder, it had six Feet, and two little Hornes, it eat *Rose leaves*, and the flowers of *Elder*.

The *Catterpillar*, the Mother of this *Beetle*, changed the First of *July*, and remained in that condition untill the 21*st.* of the same Month, then a beautifull *Butterfly* came forth which lived 9 Dayes fasting.

The birth of this Beetle , *is an odd* Phenomenon, *I am of the* Opinion *that here is a great mistake, because the like once happened to me : That* I *thought to have Observed a* Beetle *borne of a* Catterpillar *; but* I *question my owne* Observation. *The* Catterpillar *came to good ; and here is the* Butterfly *perfect. I guesse this* Beetle *might well be by accident, and unobserved brought in , and layed in the same place, where the* Catterpillar *was kept ; and so* I *formerly thought of my owne Observation :* But I *affirme nothing.* Number.

G. P. 2.
Tab. 32.

Number. 38.

The *Catterpillar* of the *38th. Table* fed of *Marigold leaves*, or *flowers* it puts off its skin foure times, and in the scarcity of Food they Eat one another; I have indeed Observed other *Catterpillars*, to Eat one the other, but those do it greedily.

After I had Fed this 14 Dayes, and that he had clensed himselfe of excrements, he changed the *9th.* of *August*, as is *Figured* in the *Table*, and after 18 Dayes, the *27th.* of *August*, came forth a *Butterfly*, having six Golden spots upon his Foure Wings: The Foure lesser Spots were upon each wing one, the 2 bigger Spots on the upper paire of wings only: This *Butterfly* lived a Hungry Life untill the *6th.* of *September*, I not knowing what to give it to eat.

We Observe in this History *that* Butterflyes, *when* Catterpillars, *are unlike themselves, not only in the disguise, but manners two; they then Eat one the other, but are neery peacefull, when in a Perfect state, aud unmaskt,* Spiders *will Eate* Spiders, &c.

I say again I cou'd have wished the Author *had taken the care of discribing to us the* Paintings *and* Colour *of the* Insects *, which he delineated with his* Pencill.

Number.

Number. 39.

The *Catterpillar*, of the 39th. *Table:* Hath his *Origin* from rotten *Willow*, and is found lying in the Bodies of those Trees.

In this Tree is found a certain fat juice, not unlike *Turpentine*, for of the Saw-duft of *Willow* a Vernis is wont to be made by boyling.

This *Catterpillar* lives of this fat juice, and is to be found in the marrow of that Tree, both Summer and Winter, for our *Dutch-Boors*, when they cleave them in Winter, often light upon them : But to have their change they muft either keep them in a warmer place, or not take them out of the Wood.

Thefe *Catterpillars* are of a redifh colour, like boiled *Cray-fifh*, and they fmell ranke.

In the Months of *June* and *July*, I have met with them creeping in the High-wayes, then feekeing out a convenient place for to change in; which choice is an old decayed *Willow*, for its foftnefs, and for its food.

Concerning the manner of the Propogation of thefe *Catterpillars*, it is obfcure, and not fully known to me, and cannot eafily be found out, for on the outfide of the Tree are no chinks or cracks, no (not fmall) holes, and yet *Catterpillars* are to be found within.

This is moft certain, that I have found thefe often in the *Pith* of *Willows*, and it is likely they are there generated of corrupt Wood by heate, as other *Animalls*, which are fpontaneoufely begot.

Alfo haveing very often tryed to keep them to know what would become of them, they all Dyed, and when I did ftop up their Dead Bodys in Juggs, I had a multitude of little Flyes, which I fay without doubt were Generated of heat and corrupted matter.

H   When

When thefe great *Catterpillars* are at their full growth and bignefs, they feek out where to reft, in order to their change, it changed in an Old and dryed *Willow Tree* the begining of *June*, as is *Figured* in the *Table*, and the 23d. of *July*, came forth a great *Butterfly*, which moved not out of its place, it was Hatched in ; and Dyed there after Eight Days , fhutting its Wings.

This very Catterpillar, *I have alfo found in the Body of* an Oake Tree, *new Feld and Sawn a funder, wherein it makes holes, you may turn your Finger in :* The Romans *had a way to feed thefe Fat, and did eat them as a delitious Food, they called them* Coffi. *It has a very rank and ftrong fmel, is a little hairy, and of a* Reddifh, *but pale colour.*

2d. The remarke of Willow *Sawduft affording a refine, is curioufe and the firft time, that I met with it in any* Author, Pliny (*lib.* 16 .c. 18.) *fayes the* Gaules *had a way of extracting a* Bitumen *from* Birtch, *which is as improbable, however to this purpofe fome of the Old Trees we dig up out of our Mountanous Moffes in the* Weft-rideing *of* York-fhire, *are certainly no other then Birtch, and when dry do Burne with as lafting a Flame, like* Firr-Tree *fplinters, which gives the occafion of calling them* Firr-wood.

3d. As to the Spontaneous *Generation of this* Catterpillar, *and other* Infects, *I have declared my opinion in the Negative.* This is moft certain, that thefe Coffi are hatched of Eggs layd by their Animall Parents, and that thefe very little Worms are capable of piercing the Tree, by little and little, that is as they Eat, that probably thefe little Holes grow up againe, after they are once fully entered, at leaft fo as not to be vifible, but to a very dilligent enquiry. Again probably they change not, but are in the Difguife of a Catterpillar, for many Years, which is agreable to my own obfervations, all which things render the Obfervation very tedious, but I little doubt, the truth of it, and that this Catterpillar,

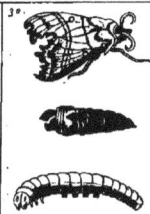

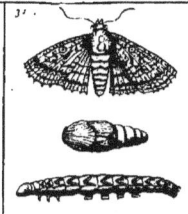

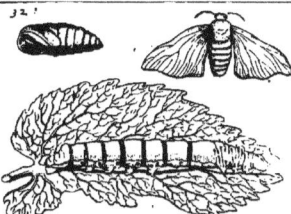

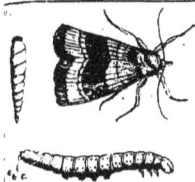

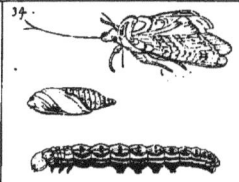

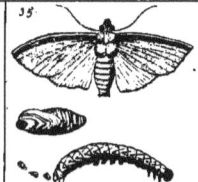

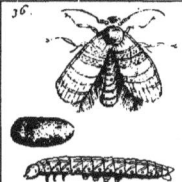

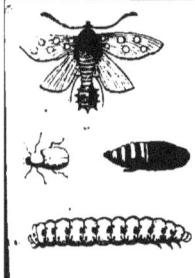

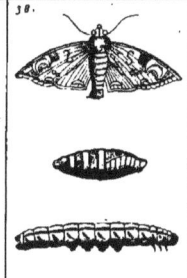

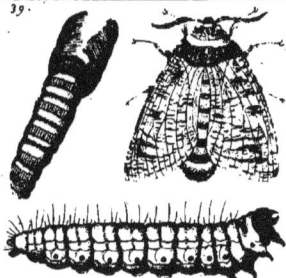

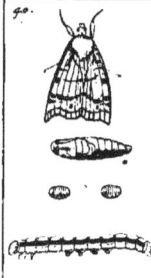

pillar *is propagated by its* Anmall parents, *the* Butterflys: *as all other Catterpillars are.*

Number. 40.

The *Catterpillar.* of the 40*th. Table.* : I took the 13*th.* of *December,* and fed it with *Sallow leaves* (its ufuall Food) as long as I could find any of them; but when I cou'd get no more of thofe leaves it fafted all Winter; but it Daicly fhefted its place, its life and motion were fo weake, as fcarcely perceiveable.

The 24*th.* of *March* , I offered it the tender Buds of *Sallow,* but it Eat not: the 2*d.* of *Aprill,* I put again before it *Sallow leaves,* and then it Eat them, for then they were bigger, and more Nourifhing; the fame thing I obferved in other *Catterpillars,* not one only; that they abftained from tender young Herbs; but eat them greedily, when a little more grown up.

The 6*th.* of *Aprill,* holding faft by the feet upon the edge of a piece of Paper, it crept out of its fkin; that I feemed to have a *Catterpillar* upon the Paper, for it was juft like (as to colour) the fkin it crept out off.

When it caft this fkin it parted with an Egg of the fame bignefs, that *Ants* Eggs are ufually of. The 14*th.* of *May,* is caft another fkin, and again parted with another Egg : Both the Eggs came to nothing.

The *Catterpillar* changed the 13*th.* of *June,* as is *Figured* , and after 10 Days came forth a *Butterfly* , which fafted Seven Days, and Dyed.

Note, that is no unufuall thing with thefe kind of Animalls; *I meane* Catterpillars, *to goe on, or fufpend their Eating, and confequently their groth, as was obferved.*

2d.

2d. *Here indeed is a perfect change, and a* By-birth *besids, I guesse the* By-birth *to be the* Chrysalis *of* Ichneumons, *because the* Catterpillar *lived: And perhaps from this Observation we may guess something of the place of* Ichneumons *Wormes feeding; which is probably not within the Body of the* Catterpillar, *but betwixt the skin or the* Exuviæ *only.*

For here every time the Catterpillar *casts a skin, it parted with an* Ichneumon, *and the reason (why these* Ichneumon *Chrysalis are* Figured *as* Foliculi, *and so first appeared: whereas others most usually* Pierce *the* Catterpillar *in the forme of* Wormes, *and change afterwards) may I say well, be the* Catterpillars *fasting all* Winter *: so that the* Catterpillar *not haeving cast all the skins to the last, these* Worms *could not break forth : Or were not fit and ripe so to do, but* Spun *themselves* Foliculi *within the* Catterpillar *, before their time, and so came to nothing.*

Thus far the smooth and not Hairy Catterpillars *whose* Histories *are more perfect, that is having both the* Disguises Figured *and the* Butterfly *two. Here follow the lesse perfect* Histories *(those in which the Figurs of the* Aurelia's *are only omitted) of the large sort of smooth* Catterpillars.

G. P.1.
Tab. 14.

Number. 41.

The *Catterpillar* of the 41 *ft. Table,* feeds on the Leaves of *Bursa Pastoris* and *Senecio* : Although most *Catterpillars* feed on the leaves of Herbs, and are therefore to be found upon them, yet this *Catterpillar* of the 41 *ft. Table,* lyes hid underground oth day time, and comes not out
untill

untill the Evening; and Eats as much ith night, as is sufficient for all the day, like *Batts.*

This *Catterpillar* changed the 28 of *Aprill,* and remained so untill the 9*th.* of *June,* when a *Butterfly* came forth which appears not in the Day, but seeks his food in Gardens amongst the Flowers.

Hiſtory. 42.

G. P, I.
Tab, 30.

The *Catterpillar,* of the 42*th.Table.* eats the Leaves of *Violets.* If it happen(when two or more are fed together) that one of them change early before the reſt, and there be a want of meat, the *Catterpillar* not yet changed will deuower the (*Aurelias*) which it may well doe, theſe being not in a condition to reſiſt and move.

This *Catterpillar* changed the 13*th.* of *October* , and continued in that condition untill the 6*th.* of *June* the year following, at which time came forth a *Butterfly* of a Blewiſh colour Figured in the *Table.*

Number. 43.

G. P. I.
Tab. 37.

Although flowers are not much infeſted with *Catterpillars,* yet are they not free from them, the *Catterpillars* of the 3*d. Table.* chiefly Eats *Gilliflowers,* I know by experiance that the *Catterpillar* lyes under ground all Day, and comes not out before Sun ſets.

He changed the 30*th.* of *July,* and continued ſo untill the 23*d.* of *September,* and then a *Butterfly* came forth of a Rediſh colour, : *Figured* in the *Table.*

Naked Catterpillars, *are a more acceptable Food to* Birds, *then ſuch as are Hairy, as I have found by experience in*

H 3 *feeding*

feeding Red-breasts : *I gueſſe the reaſon to be, that the Hair is noxious to their ſtomacks. And indeed, it is my opinion, that the* Veſicateing *faculty of* Inſects *is much more in the Haire; then in any other part: I haveing Blown into my Boxes, where ſometimes I kept a ſort of Hairy* Cimices, *had in a few Minuts after all my Face Bliſtered. Theſe Naked and therefore more inocent* Catterpillars, *by the inſtinct of Nature ſeek to preſerve themſelves, by getting under Ground in the Day time, when the Birds are ſtirring.*

G.P.1.
Tab. 56.

Number. 44.

The *Catterpillar* of the 44th. *Table*, feeds of *Dogs-mercury*, as ſoon as it perceives any thing it is not u'ſt to, it caſts it ſelfe upon the ground, and lye s round in *Ball*, as though it was Dead.——It changed the 30th. of *July*, and the 26th. of *Auguſt*, came forth a *Butterfly*, fig in the *Table*, of a rare ſhape and colour ; the fore part of its Body was Hooded as it were, it lived long without Food, and was very fearfull

What the Plant Mercurialis *may be I cannot gueſſe, many there are that bear that name, as* Bonus Henricus, Dogs *mercury*, &c.

G.P.1.
Tab. 56.

Number. 45.

The *Catterpillar*, of the 45th. *Table*, Eats the leaves of *Lovage*, only ith Night, never o'th Day-time, I found it lying in the cracks of a *Chery-tree*, it changed the ſite of its Body often in the Day time, it uſt to lift up its body, as though it looked about it watchfully.

Before

4.

5.

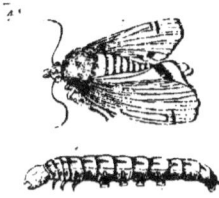

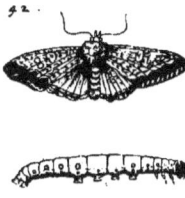

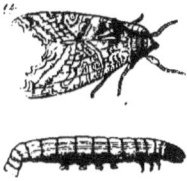

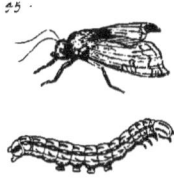

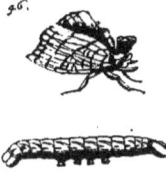

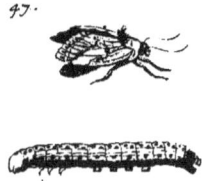

Before it changed, it caſt its ſkin, with great difficulty,
Sweating all over ſmall watery drops : having a new ſkin
it reſted for a time, and changed the firſt of *June*, and the
firſt of *July* came forth a darke coloured *Butterfly*.

Malpigius, *hath well compared the caſting of skins in*
Catterpillars, *to the breeding of Teeth in Children; becauſe
both are often accompanied with dreadfull Symptoms.*

Number. 46.

G.P. 1.
T*ab.*60.

The *Catterpillar* of the 46*th. Table*, I Fed the ſpace of
two Months, with *Ground Jvy* in *Dutch Onderhave*.
Before it began to change, It purgeth it ſelfe, and changed the 28*th.* Day of *Aprill*, and a very Beautifull *Butterfly* came forth the 26*th.* of *May*, Figured in the *Table*,
it had ſo elegant a luſter, that it cou'd not be Painted
withont Guilding, it bears two equally long Plumes upon
the Head, it is wonderfully adorned about the Eyes,
and Armed,under the Eyes appear two Teeth, it Lived
Three Days without food.

I underſtand not what our Author, *means by the two
Teeth under the Eyes of the* Butterfly ; *This is a miſtake
I gueſs :* Butterflyes, *indeed have a Tongue, Trunk, and
hollow Pipe by which they Feed on Flowers : But I am
yet ignorant of their haveing Teeth.*

Number. 47.

G. P.1.
T*ab,* 61,

The *Catterpillar*, of the 47*th.*Table : Is very curious and
delicate in its Food, which are the leaves of the *Cherry-
tree*.

It changed the 6*th.* of *June*, and the 14*th.* of the ſame
Month came forth a *Butterfly* Figured in the *Table*, of an
unpleaſant colour,and Beggar-like dreſs, it lived not long,
ſad and moving little. *This*

*This is a very quick change, the Butterfly throwing of
the disguise of an Aurelia in Eight Days, this was in the
heat of Summer, Malpigius observes, the Butterfly of the
Silk-Worm, not to doe it in the hottest season in lesse
time then 10 Days, I wou'd have it tryed whether or no it
wou'd succeed, that if a Chrysalis, change late ith Year,
was put in a warme place, immediatly after its change, it
wou'd soon throw off that disguise, and become a Butterfly,
the place might be insted of an Elaboratory, or Glasse-
house: All the parts of the Butterfly are budded in the
Chrysalis; But are not sprouted, Explicate, and hardned.*

*We come in the next place to such Historys, of the large
Smooth Catterpillars, where we find either the change imperfect, that is some by birth, Figured in stead of the Butterfly, or the Catterpillar, only, and no Aurelia, by Birth
or Butterfly at al!.*

G. P. 1.
Tab. 18.

Number. 48.

The *Catterpillar* of the 48*th. Table* : Eats *Violet-
leaves*, it can hardly endure the Sunbeames, and therefore
gets under Ground, it can hardly be found because it is
green, like the leaves it feeds on, it creeps fast enough;
save that when it perceives any thing, it is not us't to;
it moves not, but lyes as though it were Dead.

After that it had exactly cleansed it selfe of all excrements, it changed, the 4*th.* of *September*, and abode in
that condition without Meat or motion, to the 9*th.* of
May, the Year following, at which time came forth a
Fly, Figured in the *Table.*

This

This is an ordinary Flesh-fly, *and therefore the* Aurelia
was Carrion *and putrid, when the* Fly *fed upon it. To know
when a* Chrysalis *is alive, is by touching the Tayle of it, for
although it have no locall and progressive motion, while in
that state, yet it can wag that part very briskley: As to*
Aurelia's *fasting so long; as such they must fast having no
Organs to eat, and being in disguise; but whether thy are
not Fed, by a nourishment equivalent to that of Infants
in the Womb, Is to be inquired after.*

Number. 49.

The *Catterpillar* of the 49th. *Table:* I fed 13 Dayes
with the Leaves of *Willow trees,* when it began to Ab-
stain from Meat, it turned and tossed every way with
great anxiety, and labour, and then it gently reposed and
changed the 17th. Day of *July,* and the 20th. Day of
August, came forth a Yellow *Fly,* not unlike a *Bee:*
Figured in the *Table.*

We have said that Ichneumons *arc of the* Wasp kind, *it is
therefore no wonder that some sort of* Wasps *or* Bee, *neare
alike in shape to that which we vulgarly own & call* Wasps,
*shou'd be nurished after the same manner, that is, shou'd
make the bodies of* Catterpillars, *the place of their nourishment, if not the matter too, as in this* By-birth: *This is
yet a great mistery to us after what manner, and on what
these* Wasp-Wormes *feed within the Bodys of* Catterpillars; *the* Anxiety *of the* Catterpillar *in its change
argued the troublesom Guest, it had in its Body.*

I   The

Number. 50.

The *Catterpillar* of the 50th. *Table*: I for some time fed it with *Elme leaves*, when it creeps, it is twice as long, as when it rests.

It changed the 9th. of *September*, and did continue in that condition untill the 24th. of *May*, the Year following, at which time came forth an odd shaped *Fly*.

This *Fly* is a very fierce enemy to Spiders, and by a singular antipathy persues and kills them; whereas other *Flyes* are taken in Spiders webs and eaten by them: I have experienced that these kind of *Flyes*, whilst Spiders are in the middle of their Nets, and there expect the coming of *Flyes*, they seise upon the Spiders and wound them mortally; the Spider himselfe thus wounded, cast himselfe upon the Ground by a thred; this *Fly* follows, and breaks every leg of the Spider, one after the other, then he glories in a full Victory, and often goes above the Spiders body as it were rejoycing, I have Observed this thing thrice in doing, and then I saw the *Fly* carry away the Spider on its Wing.

This By-birth *is an* Ichneumon: *That is, a Wasp with a slender body: its killing the Spider is very remarkable, and I have elsewhere (in the Philosophicall Transactions) Registred an observation of these kinde of Waspes, laying their Eggs within the cakes of Spiders Eggs, and that when Hatched, the Wasp-worms do feed upon the substance of the Spiders Eggs, and do in processe of time, in the same Spiders webbs change into Aurelia's, and are thence changed into Ichneumons of their own Species, from whence we learne the confidence of these kind of Insects: Which are secure from Spiders, and therefore may well*
                                                    put

put other Infects, *such as* Catterpillars, *leſſe able to defend themſelves, to what ſervice, and uſe they pleaſe, making their Bodys* Nurſeries *of their Young.*

Number. 51.

The *Catterpillar* of the 51*ſt. Table :* Feeds of wilde *Aurach*, and is of the (*Green*) colour of the Plant; its very flow in eating and creeping, and every way a flug-gifh Animall : It caſts its ſkin like Serpents, or the Silk-worme foure times, and had four feverall fhapes, upon its caſting its ſkin, which happened thus, *viz.* When it had don eating, out of the body of the *Catterpillar* came a not-very-little worme, fomewhat flat, and fhaped like a *Fleſh maggot :* the worme crept forth the 12*th.* of *Februa-ry*, about foure in the after noon, this Worne inceſſantly crept up, and down without eating or drinking for Four-teen Hours; but in the mean time by little and little loſt its colour, and its ſkin grew harder, for defence perhaps againſt the coolneſs of the aire and enemies; and now creeping no longer, it rold its bent body, and put on an obfcure colour, and then languiſhing, it feemed to fall of eating, but not greedily : the next day it feemed to Die, there being no ſignes of life, or motion in it; but it continually reſted till the later end of *September*: At which time come forth a *Fly*, rarely to be met with; with longiſh legs, and flat feet : I believe this an *Amphibious* creature, and to live in the Water as well as in the Aire : this *Fly* had a great Head, of an unuſuall fhape, it laid mofty on its back, and therefore I have fo *Figured* him; he eat no-thing that I offered him, and died the 3*d.* of *October.*

It will be a hard matter, for any man to make ſence of this Hiſtory, *and undoubtedly the diligent* Author, *had*

much

much wronge done him, by the ignorance of the Lattin In-
terpreter, for to omit other things, what shall we make of
the XXII. Calends. October. The truth is the Animall is
very odly described, and Figured, it shou'd be a Waspe by
all circumstances, and especially in that it has 4 Wings.

Again it cou'd be no Flesh-Maggot; because the Catterpillar was not carrion; but alive, at that time of the
eruption of this Worme. The Author hints, a solution by
thinking it an Anomalous, or Amphibious Creature.
I affirme nothing, but that it was a By-birth; so the casting
of Foure Skins is false.

G.P. 1.
Tab. 62.

Number. 52.

I cou'd find nothing that this *Catterpillar* of the 52 d.
Table wou'd eat, perhaps because he had left off feeding, and was ready to change.

He began very Anxiously to tosse and roule, turning
and winding his body every way: by and by, there was an
eruption of certain drops of water or sweat out of his,
body all which drops, I saw change in the space of 12.
Hours, into living *Catterpillars* (*Erucas*) but they all
perished with their Mother, for my want of knowledge
of their food in one dayes time.

Its very frequant with our Author, to think that Dead
which sensibly moves not: I am of the opinion that this By-
birth were the wormes of an Ichneumon: And that these
Wormes after eruption changed into Aurelia's: Which was
perhaps the way they perished; compare this History,
with the Cabbage Catterpillar above, History the 7th.
of our method.

Number.

48.

49.

50.

51.

52.
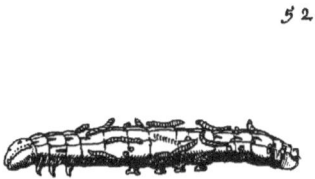

53.
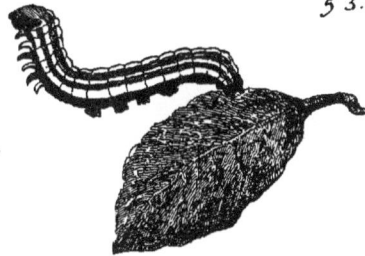

Number. 53.

G. P. 3.
Tab. T.

I fed the *Catterpillars* of the 53*d. Table*: with *Alder* and *Paretree leaves* from the 28*th.* of *May*: when it had Eaten enough, it faftned it felfe to the fides of a leafe, lifting up his head, at Sun fet it drew in its head and flept all night; but fome of thefe *Catterpillars*, will fleep 17. or 18. hours together, on the contrary others will creep up and down two Days and two NIghts without food or refting; the hinder or lower parts of this *Catterpillar* Body are pellucid, they fight defperatly together fome times; [*The Author left no further.*

We have now done with all the greater forts of Smooth *Catterpillars, as well which the* Author *left us perfect, as the more imperfect* Hiftorys *alfo. Here follow in the nex place, the leffer* Species *of fmooth* Catterpillars, *and we fhall obferve the fame order in them, which we have done in greater, that is, firft to range the more perfect and compleat Hiftorys, where we find the* Butterfly, *and both the Difguifes Figured.*

Number. 54.

G. P. 1.
Tab. 9.

The *Catterpillar*, of the 54*th. Table* Feeds of the leaves of *Ragwort:* It is a Black and Yellow colour and fmooth, in the month of *July*, it cleaves to Bents or Straws, and hides it felfe in a fkin, for matter, like Yellow Silke.

After that I had fed him with the faid herbes 8 Days, he changed the 24*th.* of *July*, and lay as dead in that condition, untill the 8*th.* of *May*, the year following, and then came forth a *Butterfly* partly Black, and partly Read, not unelegant, Figured in the *Table:* the *Butterfly*, retained

I 3

the

the colour of the *Catterpillar*, save that the Yellow in the
Catterpillar was changed into Red in the *Butterfly*, it
lived without meat 23 Dayes, and dyed after having
ejaculated its Seed.

This Catterpillar *is common and well known, I have
had some of them out of whose bodies broke forth certain
small* Ichneumon *Worms, spining very white* Foliculi.

G. P. 1.
Tab. 68.

Number. 55.

The small *Catterpillar* of the 55th. *Table :* Feeds on
Elder leaves, and roles up those leaves like a Funnell,
making it selfe a House in them against Raine, Sun, and
Birds, and for this reason also feeds only in the night time.
It changed the 22 of *November*, 1657. And continued
as is *Figured* in the *Table*, untill the 21*st.* of *July*, 1658.
Then came forth a *Butterfly*, *Figured* also in the *Table*,
very pretty and elegantly marked.

G. P. 2.
Tab. 1.

Number. 56.

The *Catterpillar* of the 56th. *Table :* Feeds upon the
leaves of Winter Roses, but will not touch any flowers
or leaves else, this is a nimble *Catterpillar*, and very sensible and crafty, I speak nothing but what I have seen
and tryed. This *Catterpillar*, (as I have often tryed)
wou'd not be moved at all, if the Winde blew the leaves
it sate upon ; but if I moved the aire with my hand it
wou'd immediatly cast it selfe down upon the ground, by
a thred of its mouth, for to save its selfe from the fall.
A mongst the leaves, it feeds of, it is wont to make it
selfe a House, elegantly knit of its owne Weaving and

in

in this it changeth, which it did the 20th. of *May*, as is *Figured* in the *Table*, and so continued untill the 12th. of *June*; then there came forth a Butterfly, which lived fasting to the first of *July*.

Number. 57.

G. P. 2.
Tab. 4.

The *Catterpillar* of the 57th. *Table*, is wont to feed of *Elme leaves*, it makes it selfe a pretty kind of Nest, fastning a thred to the one side of the leafe, and then to the other, and so brings both sides together, Knitting them close leaving an open passage at both ends; it creeps both backwards and forwards alike : If you touch it, it casts it selfe upon the ground swiftly by a thred, and do's Dart its Body swifter then an *Eale*, perhaps to strike terror into those Enemies that follow him.

I have two *Catterpillars* of the *Species*, one of them changed the 2d. of *June*, as is *Figured* in the *Table*, there came forth a *Butterfly Figured* in the *Table*, the 9th. of *July*, which lived fasting 10, Days, I guess this to be a Male the other bigger bodied *Catterpillar* lived untill the 3d. of *June*, it was full of little Wormes, which without doubt were the death of it.

As Wormes are the Death of Children, according to the opinion of both Antient and Modern *Phisitians*; So 32. Little Wormes, broke forth of the dead Body of it : The 5th. of *June*, each of which Wormes as soon as out of the Body of the *Catterpillar*, fell to work and made themselves a Net, onder which each made it selfe an Egg of its own Weaving. The 19th. of *July*, out of these Eggs came forth 32 little Flyes, which I kept alive a long time with Suggar : But at length I gave them their liberty suspecting that I shou'd else kill them, if these Fyes had been a little biggar, I had *Figured* them in the *Table*, but they were scarce visible. *These*

These Flyes were a By-birth, *and undoubtedly* Ichneumons, *becauſe they Spun, as ſoon, as they came to light.*

G. P. 2.
Tab. 7.

Number. 58.

The *Catterpillar* of the 58th. *Table*, are very noxious,
and feed of tender Roſe buds, they eat out the ſubſtance
of the Bud, and do knit the leaves together very artificially, for ſhelter againſt the Sun : This Worm is from a
fat juyce Hatched by the Sun and Dew : After long obſervation, and many uuſucſesſull tryalls (for this *Cat-*
terpillar is not eaſily brought to change by our feeding)
I found that it changed as is *Figured* in the *Table* : The
1ſt. of *June*, and was covered as with a white ſheet, the
21ſt- of *June* came forth a *Butterfly*, which as ſoon as
borne ſtood with erect wings, then it ſet a running ſwiftly with wings aloft, then it leaps up like a Graſſ-hopper,
and at laſt reſted quietly, it dyed within five Days for
want of Food.

It is ill gueſſed of our Author, *to think any thing can be*
begot of a fat juice, &c. *There is but one way, that of*
Animall Parents.

G. P. 2.
Tab. 9.

Number. 59.

The *Catterpillar* of th 59th. *Table*, feeds of Bramble
leaves, they come of Eggs, Hatched the beginning of
May,, and lived upon the dry leaves of Bramble, by the
Butterfly their Mother : For an Egg ſo layed, produced
the *Catterpillar Figured* in the *Table*, which I fed as ſoon
as Hetched with Bramble leaves, from the 6th. of *May*,
untill the 3 d. of *June*, when it began to change, making it
ſelfe

felfe a place to fecure it felfe in, of laeves knitted toge
and then changed, the next Day, as is *Figured* in
Table : The 20*th*. of *June* came forth a *Butterfly*, wlrcn
new Hatched, lay like a dead creature, but by and by
flew like an Arrow out of a Bow.

Thefe Butterflyes, *are hard to be taken abroad, being of
a fwift and cunning flight : they fit under the leaves when
weary : They abide colde well; for I have alfo found them
in the middle of Winter under Bramble leaves, this Butterfly lived* 18 *Days fafting with me.*

G.P. 2.
Tab. 12.

Number. 60.

The fmall *Catterpillar* of the 60*th*. *Table* , are ever
found almoft within the Flowers, or Bloffoms of the
Cherry, Aple or *Pare tree*, and deftroy much, and cannot
eafily be driven thence, but by fhowers, they can endure
both heate, and cold, and well know how to defend themfelves from the harme of either; they Knit and clofe up
the Bloffoms they eat, and feed moft in the coole of the
Evening or Morning, and that for 14 Days at leaft.

They are bread by a moift winde, and as foon as borne
do deftroy, and eat the faid Bloffoms like a *Gangreen;*
the watter out of which they are bread, is a moift cloud
like Honey dew, which by the fcorching of the Sun, and
the native heat of the Trees, is turned into live Wormes,
which our *Dutch Boors* call *Woolves.*

The 18*th*. of *May*, this *Catterpiller* changed, and the
18*th*. of *June* came forth a *Butterfly*, quiet for two hours
after it was borne: which as foon at its wings were dry,
let fall one drop of water and flew away.

In the Morning you will finde of thefe *Butterflys*, fitting on *Paretree* flowers , and fucking Dew thence, but
in Winter they hide themfelves in Stables, and Grainaries

K.   for

for warmth, they are very fearfull, and scarce fly away unless much urged : This *Catterpillar* was of old called by the *Dutch Boors : De Woolfe.*

Our Author here also follows the vulgar opinion, as well as name, concerning the breed of these Wormes ; but his own Observation of the surviving of these Butterflys all Winter, were enough to evince the contrary, that they breed the Wormes : Compendious expedients to rid a Fruit Tree of them, were well worth the invention ; in the meane time, it is as well worth our pains, the picking these out of Blossoms betimes, as the weeding of Corne is. And my reason is, because they are bred but once in a season or about a time ; and therefore if you rid the Fruit once of them, the Fruit is secure of them ever after. But of these I purpose to say more in an other place.

G. P. 3.
Tab. 13.

Number. 61.

The *Catterpillar* of the 61*st. Table*, feeds only on the Herbe *Calamintha*, it is alwayes wet, and leaves a shineing slimy tract after it, where it has crept like a Snaile : its a fearfull and timerous *Animall*, and hides it selfe under the leaves of Mint, and often changes its station, it feeds ith night, but never ith Day time, and with its body in an erect posture, as one that is very circumspect and wattchfull to avoid danger : when it goes, it makes hast.

The *Titmise*, or little *Birds* (*Pari*) devower these *Insects*, and therefore they seldom appear, and are not to be driven from the places they hide themselves in, but by violence.

The *Catterpillar* Figured in the *Table*, abstained from meat the 15*th.* of *October*, and made it selfe a little house of sand and slime, and loosing its colour it changed the 26*th.* of *March*, as is *Figured* in the *Table*, and the 13*th.* of *June*, came forth a *Butterfly* very notably marked with spots, which after 8 Days fasting dyed.

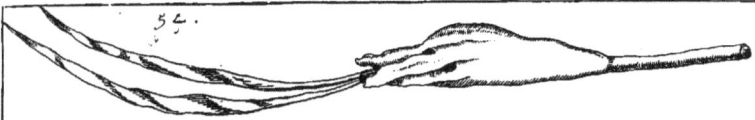

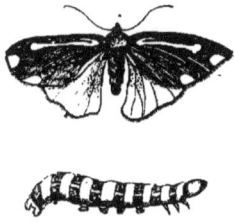

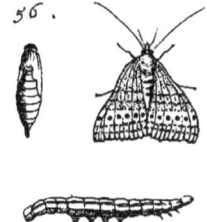

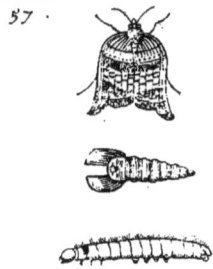

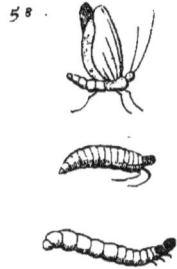

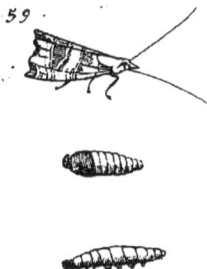

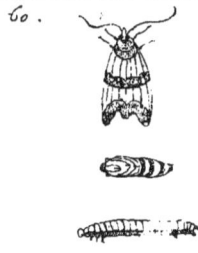

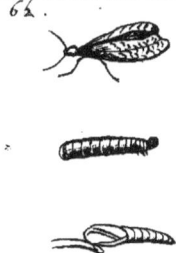

I conceive there is but little difference betwixt the Saliva of Infects, *and that of* Spiders, Snailes, *and* Catterpillars, *being very much akin. The reason why the* Spiders *and* Catterpillars, *forme the Thread of their Saliva, and the* Snail *not, is very probably, the Organ or Pipes, through which this juice passes, which are not given to the* Snaile.

Number, 62.

The *Catterpillar* of the 62. *Table,* fed of the putrified *Stone* of a Mountaine *Duck,* and was bread thence, and lived in that ftone as long as any thing remained to feed on.

It change the 29th. of *May,* 1659. As is *Figured* in the *Table,* they undergo the like change in the Fethers of *Duck,* wrapt up and hid therein, that nothing but a little hole was left for the *Butterfly* to go out of.

The 7th. Day of *June,* 1659. came forth a *Butterfly,* or Moth *Figured* in the *Table,* this Moth is beautifull and delights in flying, wherein it ufes ftrange motions and windings. Thefe Moths live long unleffe they be taken in Spiders webs.

They fhun the light and hide themfelves in obfcure places, and fometimes in Gardens, under the leaves of Plants: They feed on the fweet Dews on Flowers, as other Flyes; they hide themfelves in winter in Houfes, &c.

From this *Hiftory,* fome have doubted whether the *Catterpillar* was ever perfectly formed before its change, but was rather borne after the change, becaufe that untill the change it remained in the place of its birth, and took its nourifhment there as Children do in the Womb, and Chickens in their Eggs.

The Butterfly *living abroad in Summer, and returning into Houfes in Winter, is enough to fatisfy the manner of this* Catterpillars *birth, which is from the* Animall *Parent,*

K 2   *and*

and not of putrified flesh, also it was in disguise in the rotten flesh, not as in the Womb, but as in a place where store of food was to be had; sutable to its nature, and for this the Mother Butterfly was Caterer; so that however the parts of the Butterfly are altered, from what they were, when in the disguise of a Catterpillar, yet do's the Butterfly retaine a sense of that meat, which she once fed off her selfe and do's not carelesly drop her Eggs; but there only where she finds sutable food for them to eat, as soon as they shall be hatched.

Here follows the less perfect Histories, of the lesser Species of smooth Catterpillars; That is to say; where either one or the other of the Disguises are not Figured in the Table, or a By-birth only instead of the Butterfly.

History. 63.

G P. 1.
Tab. 15.

The Catterpillar of the 63d. Table feeds on Thorne tree leaves as long as there are any to be had, it goes backwards aud fore wards, as it pleases and that swiftly, when these Catterpillors, change, they get together into oue place, and every one hangs by his own proper thread.

This Catterpillar, chauged by little and little, the 5th. of June, and the 27th. of the same Month, came forth a Moth Figured in the Table, which Moth haunts Thornes, and layes it Seed upon them which (spawn) hangs thereon untill the begining of the Year following, for Catterpillars appear not till there be food for them.

I have seen the change of this Catterpillar, (see my Notes.)
Numb-

Number. 64.

The *Catterpillar* of the *64th. Table*, feeds of Sallow leaves, they shelter themselves from the Sun, in boughs and leaves knit together, they creep swiftly, and as soon as they perceive anything that is not usuall, and not familiar to them, they suddenly cast themselves down by a thred for their security.

This *Catterpillar* changed the 53th. of *June*, and a nimble *Butterfly* Figured in the *Table* broke forth the 21st. of the same month.

Number. 65.

The *Catterpillar* of the 65th *Table*, feeds of the leaves of the *Ashtree* (which are Infested by this and many other *Animalls*) it very cunningly rowles up the leaves, and shelters it selfe against the Sun.

This *Catterpillar* purged it selfe and changed the 24th. of *June*, and the 26 of *September*, came forth a Moth Figured in the *Table*.

Number. 66.

The *Catterpillar* of the 66th. *Table*, was bread in a *Sugar Pear*, and feeds thereof, it *changed* the 3d. of *August*, and ths Second of *July*, the Year following, that is, after a 11 Months (in which space it lay without motion or food , and as 1 thing dead) came forth a *Butterfly*, *Figured* in the *Table*.

We must think that Fruits, as well as leaves, are not the Equivocall Parents of any Insects : *But the Butterfly only of* Catterpillars, *&c.*   K 3   *Not*

Not withstanding the opinion of Aristotle *(* de Plantis *)*
and lately of Signior Read, *concerning Plantigenous*
Animals.

In the next place we shall range the Historys *of the lesser*
smooth Catterpillars, *which are imperfect, that is where*
there is a By-birth *only, and no* Butterfly Figured.

---

7. P. 3.
ab. B.

Number. 67.

The *Catterpillar,* of the *67th. Table,* greedily Eats the
the leaves of *Sallows,* there are great numbers of them :
Also these *Catterpillars* drink much and longe, especially
any sweet things: If you touch them, they defend themselves moving very swiftly the hinder part of their body.
this *Catterpillar* changed the 1cth. of *Septembre,* 1663.
and continued so untill the 22 of *August,* 1664. And then
came forth a *Fly,* which lived fasting untill the 30th. of
the same month.

In the very middle of the change, is seen a black Egg,
to which this *Fly* owns its birth.

This is the By-birth *of some* Flesh-fly, *and our* Author
is very diligent in that he has here, once for all, observed to
us, that he found, in the very middle of the Chrysalis *of*
the Eruca, *the* Chrysalis *of the* Fly ; *which he calls a black*
Egg : *The* Catterpillars Chrysalis *was undoubtedly Car-*
rion, and then Fly-blown.

Number.

Number. 68.

G. P. 1.
Tab. 28.

The *Catterpillar* of the 68*th*. *Table*, Eates the leaves of *Roses*, and I feed him also with the leaves of *Province Roses*, as often as I neglected to feed him, he prepared for change, but because that I had often observed that from an untimely change, ugly, miserable, and imperfect *Butterflys* came, therefore I gave this meat as long as he woud eat, and that he willingly changed : He clensed himselfe from all excrements, and changed the 14*th*. of *August*, and the 17*th*. of *June*, the Year following came forth a Yellow Fly, which is *Figured* in the *Table*, this Fly was slow and weake : This *Catterpillar* lay in the change, more then 10 Months, like a dead thing without motion and feeding.

Number. 69.

G. P. 1.
Tab. 29.

The *Catterpillar* of the 69*th*. *Table*, delights to feed of the leaves of the Red *curran-tree* : That only when the Day breaks, after having quickly cleansed its body, it began to change the 12 of *July*, and continued in that condition untill the 14 of *August*, at which time came forth a Fly, *Figured* in the *Table* : A very swift Flyer but Dyed within Foure Days.

These two last are By-births, *and of the Waspkind, being something more grosse and thick bodyed, then we usually find* Ichneumons *to be : however they are of kin, and I little question, but our Author might have found in these* Chrysalis's *the Ichneumons* Chrysalis's; *or the Nimph's of Wasps enclosed, if he had had the hap to have sought for them.*
By the Figure of the Catterpillars *they seem to be of one*
                                                     and

and the same Species, *and our* Author *not distinguishing betwixt* By-births *and such as are* Genuine, *he seems in some few places, to have multiplied* Species *unnecessarily.*

Number. 70. A.

The *Catterpillar*, of the 70 A. *Table*, feeds on the leaves of the *Elme*, and is to be found on the very top of these trees: it leaps from place to place, like a *Graßhoper:* cold destroys it.

Having clensed its body, it *changed* the 28th. of *August*, within a net which it had made before its *changed*, and which was as bright as Siluer, it continued in the change till the first of *June* the Year following, about 9 Months, at that time came forth a Fly *Figured* in the *Table*.

This By-birth, *is a* Waspe, *as appeares by the* Figure.

Number. 70. B.

This *Catterpillars* feeds of *Elme* leaves, their sleep extending 10 hours, and then cleanse and eat againe, when they are about to change, they let themselves down from the Tree by a Thread for safety, as soon as they come to the ground they seek a dry *Elme* leaf and having found one they creep into it, and Artificially Knitting it together clse on all sides, covering their bodies besides with a silver coloured and bright Net: In this manner changed the *Catterpillars*, the 26 of *May*, as is *Figured* in the *Table*, and Foure Days after came forth a Fly, *Figured* also to the life in the *Table*: It lived with me also fasting Six Dayes.

This

This is an other History, *of the same* Catterpillar, *and this is a* By-b.rth *also; but a Flesh-fly, and Viviparous, if I mistake not.*

Number. 71.

G. P. 1. *Tab.* 48.

The Catterpillar of the 7 1st. *Table.* feeds of the leaves of the Cherry-tree, they are for the most part to be found under the leaves, shaded from the Sun.

This *Catterpillar* cleansed it selfe, and changed the 22*d.* of *July*, and the 4*th.* of *August* came forth a Fly.

This By-birth (*by the* Figure*)is a* Waspe; *having a short, thick body, and* Foure Wings.

Number. 72.

G. P. 1. *Tab.* 49.

The *Catterpillar* of the 7 2*d.* *Table.* eats the tender leaves of the Ash, leaving nothing but the skins; it is of a shining colour, as black as Pitch; it grows no bigger, then it is. *Figured:* I called this the black Crab, because its hinder parts are like a Crab.

It changed the 28*th.* of *July*, and there came forth a Fly, *Figured* also in the *Table.*

This By-birth *is a* Flesh-fly, *as appears by the* Figure; *and therefore we suppose the* Chrysolys *was* Carrion.

Number. 73.

G. P. 2. *Tab.* 2.

The little *catterpillar*, of the 7 3*th.* *Table.* eats the green leaves of Worm-wood; and its bred of a certain hn-

L   mour,

mour, as a feed, which some Mother Fly laid upon that
Plant, Whence comes this *Catterpillar*: When it is newly
hatched, it is of a white colour: When it begins to creep
upon the ground, it forthwith makes it selfe a shelter, a-
gainst the scorching heat of the Sun: After that it has
once attained to its full growth, it is incomparable swift,
and not easily to be taken, at the least noise it runs
out of its shelter or lurking place to escape the Birds, it
is then somthing like the colour of the Earth, it falls on.

When it begins to change, it gnaws in two a sprig of
Wormwood, that it may the more conveniently knit it selfe
therein a House, to change in.

This *Catterpillar*, changed the 4th. of *June*, as is *Figured* in the *Table*; the 9th. of the same Month came forth a
Worme, which the 13th. of the same Month appeared in the
shape of an *Egg*; out of which *Egg* came a *Fly* (Figured in
the *Table*) The 26th. of the same Month; which in less
then a quarter of an hour, became as big again, as when
first hatched: I nourished this *Fly* along time with Suggar and Water; it made a wonderfull loud noise in flying,
though but little of body.

*This By birth, is a Flesh-fly, which are well known to
fill a Room with noise, when they fly about it: Here are
some unintelligible mistakes in the* Latin *Coppy.*

Number. 74.

The *Catterpillar*, of the 74th. *Table*, eats the Leaves of
the *Rose tree*; for the most part it feeds in the night only,
for fear of *Birds* perhaps: It creeps slowly, and if touched, it slowly rowles it selfe up: When it is full, it
stretches it self out as though it did, it makes it selfe a
House to change in, like a half Net, very pretty.

It changed therein (as is *Figured* in the *Table*) the 16th. of *September* : And the 14th. of *May*, the Year following came forth a *Black Fly*, which lived 7 Days fasting.

The By-birth *here is an* Ichneumon, *as appeares by the* Figure.

Number. 75.

G. P. 2.
Tab. 5.

The *Catterpillar* of the 75th. *Table*, would eat nothing but a certain *Rush* or *Grasse*, growing on the Banks of Ditches : It fed but flowly, and crept as flowly after it had done feeding, and when it was full, it was a third part lefs, then in the time of its feeding.

The 2d. of *June*, came forth of this *Catterpillar*, a *Worm*, out of which *Worm*, the 9th. of the fame Month came a *Fly*, which was fo very fmall, that I could not *Figure* it in the *Table*.

This *Catterpillar* changed in the forme of a *Tobacco-Roll*, as is *Figured* in the *Table*.

The 1ft. of *Auguft*, came forth thence a *Fly*, whofe wings were fo clear and *Tranfparent*, and clapt fo clofe to its body, that it feemed to be without wings : This *Fly* was like a great *Pifmire*, and ran fwiftly, and not with Expanded wings, but with them clapt clofe to its body, it lived with me fafting 14 Days.

'Tis rare that one Catterpillar fhould nourifh two By-births: but this Hiftory, feems to be an inftance, that it is fo, and that both of them are Ichneumons.

Number. 76.

G. P. 2.
Tab. 6.

The fmall *Catterpillar* of the 76th. *Table* eats *Columbine* leaves, and cleaves fo faft to the edges of thofe leaves,

L 2   that

that you may sooner pull them in pieces, then take them off: they eat like a spreading *Cancer*, what ever Plant, be it *Roses*, &c. they light on; for the most part you find them great and small together; they are just of the colour of *Columbine leaves* : they fear no weather : touch them and they vibrate their tail wonderfully.

This *Catterpillar*, changed the 8th. of *June*, as *Figured* in the *Table* : And the 21*st.* of *June*, came forth a Black *Fly*, Figured also in the *Table*; these kind of *Flyes* I have often seen sitting upon Bryonie, (*Vuæ urſinæ*) lived two Days without meat.

This By-birth *is a kind of* Flesh-fly; *and the* Chryſaliſ *is the change of a* Flesh-fly ; *and therefore here is ſome eſcape in the obſervation.*

G.P. 1.
Tab. 64.

Number. 77. *a.*

The *Catterpillar* of the 77*th. Table*, Marked *a.* eats but once a Day, and that ſparingly, the leaves of *Sallow*, and when it has done eating, it rowles up its body in a round, like a (*Snake*) dog, and then about noon the next day a little more food, and then to reſt in the ſaid poſture : this I obſerved it to doe (and it is a moſt wonderfull *Hiſtory*) untill the laſt day of *September*, 1653. and it reſted without *changing* the poſture of its Body, or moving untill the 24*th.* of *October*, 1655. I touched its body daily with a little Feather, that I might certainly know whether it was a live or no, and I obſerved it alive, and at every light touch to move and contract its body, two whole Years and 24 Dayes, and that without any food, or ſo much as locall motion, I ſaw no *change* in it, but that its body through long faſting was manifeſtly leſſened.   Num-

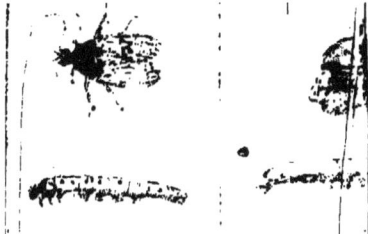

6 A.

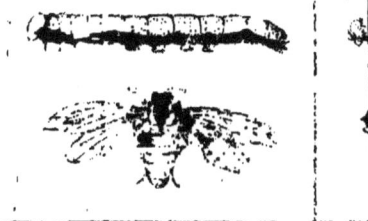

2.

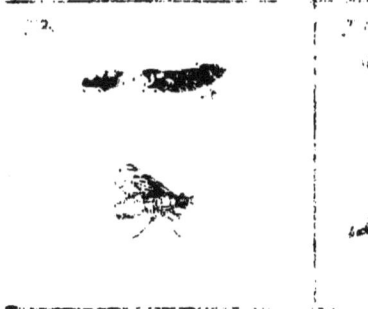

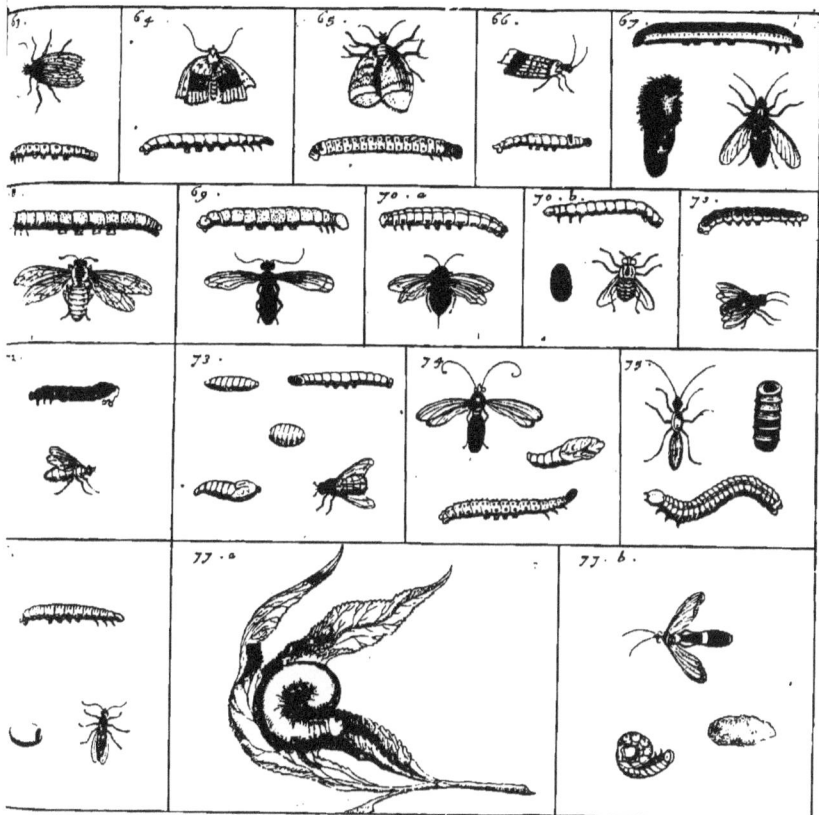

63. 64. 65. 66. 67.

68. 69. 70.a 70.b. 71.

72. 73. 74. 75.

76. 77.a 77.b.

Number. 77. b.

The *Catterpillar* in this 77*th. Table*, fed for most part on *Columbine*, and some times on *Rose leaves*; when full, it roled it selfe up like a Hedg-hog, as is *Figured* : I kept some of these *Catterpillars*, some whole Years, formerly by me, and fed them, but they constantly dyed without changing : without doubt, because somthing was wanting necessary to their change; which thing when I had thought on, I shut one of them up in a Glass Violl, filled with Earth and fed it therein, after a while, I found that it crept within the Earth, and that he made himselfe a large House in the bottom of the Glass, by the slime of of his mouth, Arched over head, and covered himselfe with Earth, that he seemed a lump of Clay, and thus He did, as I guess, for safety : Also within this Earth, he knit himselfe a Net for yet greater safety.

The 27*th.* of *July*, this *Catterpillar* rested in order to its change, as is *Figured* in the *Table*, and remained therein 7 Days without stirring, the 5*th.* of *August*, came forth a *Fly*, and the day after another *Fly*, both which after they had rested a little while, did couple in order to the propagation of their *Species*.

I have often seen these *Flys* lay their seed upon the leaves of *Columbines*, which is a Green and fat juice, and so small, that one must have good Eyes who discernes it.

These *Flys* I fed for some time with Water and Hony, at length depriving them on purpose, of their Water, they Dyed.

This is the 3d. time our Author has made two Histo-tys of one Animall.

Here are indeed many things worth nothing, and diffi-

L 3   culty

(78)

cultly explainable: In the first History, the long suspending
its Change, even for above two Years, is very remarkable.
In the Second History, the diligence of our Author, in
supplying it with Earth is very commendable.

Again the two By-births are Ichneumons, as appears
by the Figure: Their coupling is extraordinary; the Author, never yet having observed it; though some Scores have
been born together.

He also tells us of the Seed or Eggs of these Ichneumons,
which he sayes (from frequent observation) is extreamly
small, not to be seen, but by very good Eyes, but that I suspect
an errour in this observation, (this fat and Green juice,
may rather be an excrement then Eggs) I shou'd think that
possible the Ichneumons Flys Eggs, may be licked up, and
swallowed down by Catterpillars in feeding, and escape
digestion; and hatch within the Catterpillars body: but I
affirme nothing not having yet seen the Seed, or Eggs of any
one Ichneumon layd.

Hitherto of the smooth Catterpillars, both Small and
Great: In the next place we shall Range the Historys of the
Hairy Catterpillars.

G. P. 2.
Tab. 39.

Number. 78. a.

The Catterpillar of the 78th. Table, marked a. feeds on
Alder leaves, I nourished two of them, to see what wou'd
come of them; the one was a little brighter coloured, then
other.

One

One of the two *changed* as is *Figured* in the *Table*, the 15*th*. of *June*, and the 15*th*. of *July*, came forth a *Butterfly Figured* in the *Table*, very brisk and sportive, I kept him alive 12 Days.

I had a Third of these kind of *Catterpillars*, which seemed to be sick, it had no stomach to its meat, and its *Hair* trembled, as though it shook with an Ague.

It notwithstanding changed (just as the first is *Figured* to have changed) the 20*th*. of *June*, and out of that change came forth a *Fly* the last Day of *June*; *Figured* also in the *Table*: This *Fly* was very nimble in running, and very sportive with its Wings and Hornes, it eat Honey exceedingly, and immoderatly, I let it goe, now well knowing to what it ows its *Birth*, and what comes of its Seed layed upon *Alder leaves*, and Hatched by the heat of the Sun.

Number 78. *b.*

The *Catterpillar*, Companion to the first, (and of which we said came of *Butterflys Figured* in the *Table*,) changed the 19*th*. a *June*, just as *Figured* of the *change* of the first.

The 2*d.* of *July*, came forth a wonderfull *Animall*, without Wings, having the hinder part of its body very thick, and full of Eggs, and covered with a thin skin, so that the Eggs were almost visible through it, the Day after its Birth, it layed all its Eggs, and having that thin skin of its Tail, ful of Down or Wool, like the *leaves* of *Moth Mullein*, it pull'd of that Down, and covered its Eggs with it for security, and then when it had emptied it self, there was little left it, but the R repart of its Body and Leggs, and it appeared then much like a *Spider*: it lived fasting 18 Days. The very day that it Dyed the Eggs were hatched and the little *Catterpillars* crept about, I suffered them to provide food for themselves, not

being

being willing to be troubled with the feeding of them, especially knowing what they wou'd come to.

Note that all that is related in the 1st. History, is agreeable to our observations, there being an Ichneumon from the Chrysalis, of an Eruca: also besides the By-birth, there is the Butterfly, viz. The Genuin and Legitimate offspring.

In the 2d. History, there are two things very singular: The First is the birth of a Monster, or a Butterfly, without Wings, this I say is no By-birth, but a Monster in Nature, such as the birth of a Bird wou'd be without Wings. The Second is yet more wonderfull; that this she Monster should lay her Eggs (which thing is common to all the Butterfly kind as soon as born) without the Copulation of a Male, and that (which is most remarkable) these Eggs shou'd be Hatched, that is Prolifick, without the Male, (Vid. Malpigium de Bombyce) where one of the best observations in that piece is the change of colour in the Eggs, of such Females as have admitted the Male; and that such Eggs as are layd without the admittance of the Male, doe not change into that colour, and are not Prolifick; This I can affirme that I had a Butterfly which layd her Eggs in a Box, and that these Hatched therein after a certain time, but whether the Butterfly had admitted the Male before I took her, I cannot say.

G. P. I.
Tab. 59.

Number. 79.

The *Catterpillar* of the 79th. *Table*, feeds upon *Plumtree leaves*, it is a wonderfull *Figure*, on the back it has Four Yellowish Brushes, or Tufts of hair, out of the Fore-head grow two other Tufts, like Snail hornes, on each side are two other Tufts like *Oars*, the one Black,

the

the other Yellow, on the hinder part of the Body they look like Feathers; but all is Hairy; it *changes* its skin with great anxiety, wiping its sweaty Body with the Fether-like tufts: all the Day, after the shifting of its skin, it rests without food; and all its body is very tender and soft.

After it had cleansed it self well, it change the 20th. of *June*, and abode in it untill the 30th. of the same Month, and then came forth a wretched creature, neither *Butterfly*, nor *Catterpillar*: the reason of the defect was, that it abstained from meat before its time, its Body not being arrived to that perfection, requisite to its change.

The Authors *words must be favourably interpreted, for it is plaine in the Table, that the* Animall *was a Butterfly; but as he well notes a starved thing; many of which I have had, whose Wings were yet imperfect, or at least not in a condition to be expanded.*

Number. 80.

The *Catterpillar* of the 80th. *Table*, is found about the Sand hills, along the Sea shore, where it feeds of various Herbs but I could not find one Herb that it wou'd eat off: upon the back of it grows Five Bunches of hair, two upon the head like Hornes, and one upon the Taile; and out of these Bunches grows some Haires longer then the rest, as ornaments: I forthwith designed this *Catterpillar*, because it hastened to *changed*, making a web for it self of its own Hair.

It changed the 10th. of *July*, and the 8th. of *August* came forth a *Butterfly*, which as soon as borne hid (as it were) it's eyes with its fore feet, as not being yet able, to endure the light.

M   Number.

G. P. 3.
Tab. S.

The *Catterpillar* , of the 81*ft*. *Table*, was taken in an *Apricock* tree : And fent to me from *Bergenopſone*, *&c*. After I received it, it wou'd not feed, but changed the 7*th*. of *October*, 1664. And the 2*d*. of *May* the year following came forth a gray *Butterfly*, which ſo covered with its Fore-feet the whole Head, that it was not to be ſeen. It ſeemed to be a Female from the heavineſs and thicneſſe of its body : It lived faſting 10 Dayes, and before it dyed, it layed many Eggs ; but which for want of the Seed of the Male were not *Fæcund*.

How much this opinion, of Infæcond *Eggs agrees with the Obſervation of the 7* 8*th*. Hiſtory , *Let others Judge.*

G. P. 1.
Tab. 50.

Number. 82.

The *Catterpillar*, of the 28*th*. *Table*, feeds only of a certain Graſſe ; it ſleeps in the Day time, but ith night is in perpetuall motion, its very thirſty, when it drinks, it dipps its Head in the water, and lifting up the forepart of its body, it drinks like a Hen, often ſipping.

When it had cleanſed it ſelf, it changed the 7*th*. of *June*, and abode ſo untill the 3*d*. of *July*, and then came forth a *Butterfly* ; a pretty big one, of a yellowiſh colour, marked with a red Line croſs each Wing, as *Figured* in the *Table*.

G. P. 3.
Tab. G.

Number. 83.

The *Catterpillar* of the 83*th*. *Table*, wou'd eat nothing that I gave it ; it is a ſlow goer, not unlike that ſlow
paced

paced *Animall* , well known to the *West Indians,* by the Name *Haut:* It was wont to wipe and cleanfe its whole body, with a dark coloured Feather on its taile, and likewife defend it felf with the fame, if touched: It is very quiet, before its change; it *changed* the 2d. of *October,* 1663.

It compofed it felf for change thus : It firft made a Net, round about its body, of its own juice ; and within that Net another, yet leffe and of a finer thread; which touched not the outward Net, but hung like a Bell in it, and was fo thick and compact that no body cou'd fee through it; and within this it changed.

The fecond of *May,* 1664. Came forth a lufty *White Butterfly,* I ftuck him throw with a pinn, that he might not Diffipate its excellent Whitneffe, and fo he lived 4 Days.

Number. 84.

G. P. 2.
Tab. 50.

The *Ruff Catterpillars,* of the 84*th.* *Table.* Eats greedily the green leaves of the *Artichoke* plants; and when its belly is full, it creeps into the Earth.

This *Catterpillar* put off its fkin the 30*th.* of *Auguft:* and fafted, until the Haire was grown again, and that it cou'd eat ; that is for foure or five Days, fooner or latter.

The 9*th.* of *September,* it changed, as is Figured in the *Table.*

The 1*ft.* of *May* the Year following, came forth a *White Butterfly;* which never moved for two Days ; but then grew wild and furious; and at length dyed for want of Food.

Number. 85.

G. P. 3.
Tab. H.

The *Catterpillar,* of the 85*th.* *Table,* eats *Alder leaves* ; all its body is Rough and Hairy; if you touch thefe

M 2   Hairs

Hairs, it Vehemently ftrikes its Head with its Tail; in one Night it loft all its greeneffe, and became wholy Black : I cou'd well fee a *change* of colour with my Eyes ; it fhook the Haires and moved them with a notable force, and cleerly without *changing* at all its fkin.

It changed the *6th*. of *October*, 1663. As is Figured in the *Table*.

The *30th*. of *June*, the Year following, came forth a duſkiſh coloured *Butterfly*, marked with white fpots, as is *Figured* in the Table : It lived untill the *10th*. of *July*.

Number. 86.

This *Catterpillar*, of the 86*th*. *Table*. creeps fwiftly from place to place, and eats almoft any kinde of Herbs, *Trefoile*, *wild-time*, *Elme*, and *Willow leaves*, and many others ; and yet moft other *Catterpillars* will ftarve before they will feed of any thing, but their ordinary food : For colour it was of a fhining jett Black : if touched; it rould it felf up round like a Ball.

It changed the 13*th*. of *September*, 1963. And the 25*th*. of *May*, the Year following, came forth a very beautifull White *Butterfly*, having its body diftinguiſhed, with three rows of Black Spots ; it lay without motion.

It is worth Noting, that from a Cole black *Catterpillar* that had no White, but about its mouth, there fhou'd be born a *White Butterfly*.

From another of the fame *Species* of *Catterpillar*, the 13*th*. of *June*, 1664. A *Long Legged Fly* came forth, which lived not above one Day.

Number. 87.

The *Catterpillar* of the 78*th*. *Table*, feeds on *Poplar*, and *Willow leaves*, which two are of a like nature: Thefe *Catterpillars*

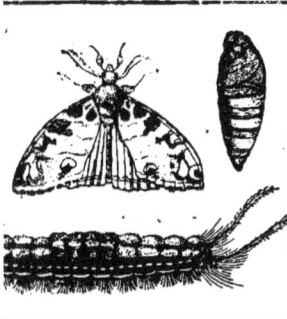

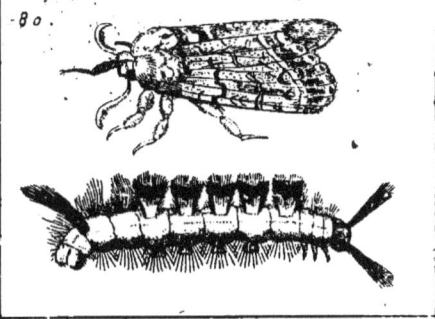

80

84

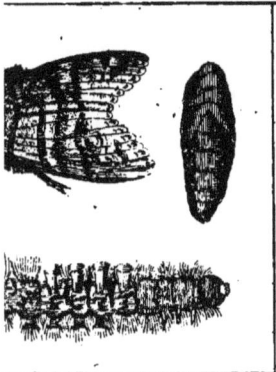

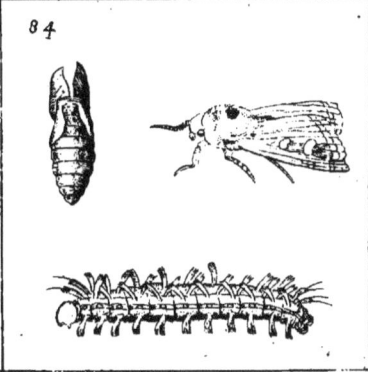

86

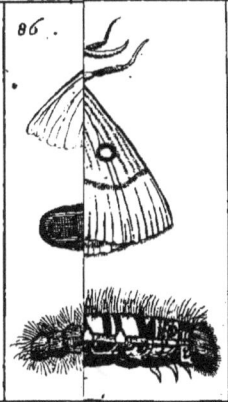

89

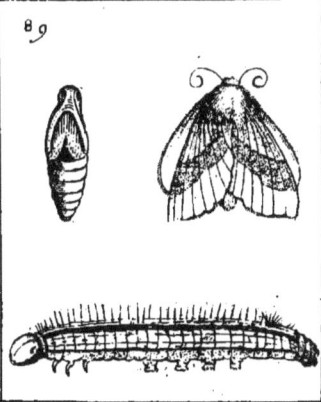

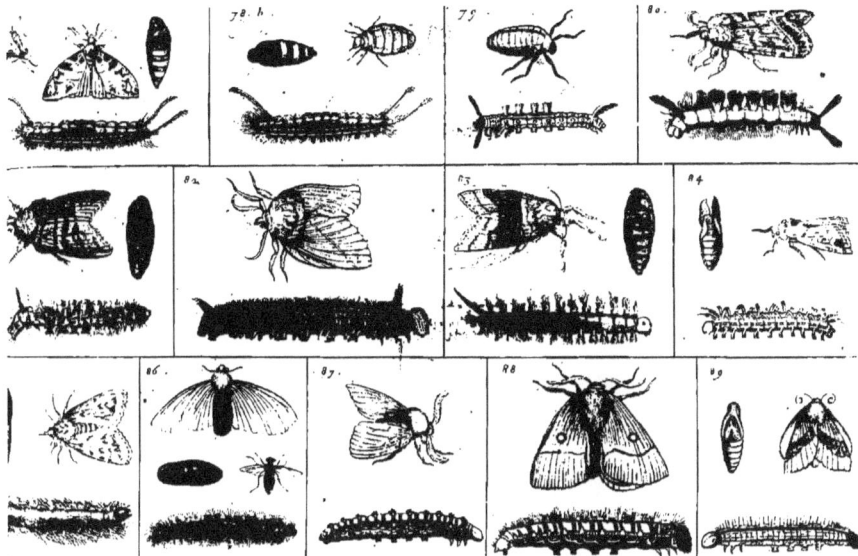

pillars come of a Seed, as moſt do, which did ſtick to the *Poplar leaves,* they are ſeen a far off, abiding in the tops of thoſe Trees, becauſe of their notable whiteneſs, *Figured* in the *Table.*

It changed the *6th.* of *June,* and continued ſo 14 days, and then came forth a *White Butterflys* Figured in the *Table.*

The *Butterfly,* laid its Seed after ſome Dayes time, and lived, without food, 24 Days.　　　　　　　G. P. 1. Tab. 7.

Number. 88.

The *Catterpillar* of the *88th. Table,* Feeds upon the leaves of *Brambles* and *Oziers.*

It changed the 13*th.* of *June,* and ſo remained untill the 14*th.* of *July,* and then came forth a dark coloured *Butterfly,* Figured in the *Table :* It is of a browniſh colour, upon each wing it has a round white ſpot ; I found it dead after two Days, though it ſeem'd to be of a ſtrong nature.

Number. 89.

G. P. 1. T. 10.

The *Catterpillar* of the 89*th. Table,* feeds of *Sallow leaves;* which are dry and aſtringent, and therefore it drinks much ; which are ſingular to it ; for almoſt all others eat green leaves, and drink not.

Its birth is from a Ring of little Eggs, cleaving ſo Tenatiouſly to the tender twiggs of trees, that it can ſcarce be gotten off, but with a Knife ; theſe Eggs abide the ſharpeſt Winters, and are hatched by the heat of the Sun in the Spring, when the leaves put forth, that their meat may be ready for them, as ſoon as born.

This *Catterpillar* changed the 2d. of *June,* and continued ſo untill the firſt of *July,* ; and then came forth a *Butterfly* Figured in the *Table.*

It dyed after it had layd its Eggs ; but the Eggs dryed, for want of having been beſprinkled with the Seed of the Male.　　　　　M 3　　　　　Num-

Observe the Elegant Posture of the laying of the Eggs, of this Butterfly ; in the fashion of a Ring, Circling a Twig ; as for their being in fæcund, for want of the Male, it agrees not with other, this Authors Observations, as is above noted.

G. P. 1.
Tab. 13.

Number. 90.

The *Catterpillar*, of the 90*th. Table*, feeds on *Haw-thorn leaves*, and *Pear-tree leaves* : As soon as *Autumne* comes, they gather together, and *Club for a Web* ; for that purpose they knit together the tops of the yet. tender boughs with their leaves ; and therein preserve themselves, as soon as the weather is favorable to them, that is about the beginning of *Aprill*, they begin to eat againe ; they leave a *Hole* open, out of which they may goe and come in a-gaine ; in cold weather they exactly shut that *Hole*: *Rain* never stands upon the web, but slides off, as though it was greasie : *Titt-birds*, eats them greedily.

This *Catterpillar*, changed the 2*d.* of *June*, and the 30*th.* of the same Month came forth a White *Butterfly* marked with a Red spot on each wing.

G. P. 1.
Tab. 13.

Number. 91. *b.*

The *Catterpillar*, of the 91*st. Table*, eats *Haw-tree*, and *Cherry-tree leaves*, &c. It is wont to ly under the shade of the leaves ; as hardly induring the heat of the Sun.

Before it changed, it cast its skin thrice, each time, rest-ing on eay until its Head hardened, for immediatly after its skin casting, the Head, and all the Body is soft, and weak ; and by little and little grows stronger and hardens.

It *changed* the 12*th.* of *June*, enclosed in a Bagg, as is *Figured* in the *Table*,: And the 4*th.* of *July*, came forth an elegant *White Butterfly, Figured* in the *Table*.

Number.

Number. 92.

The *Catterpillar* of the 92 d. *Table,* feeds on the leaves of *Cherry, Pear, Plums,* and *Almond trees* : I could never find the feed it came off: It is admirably beautifull in the variety of colours ; when it is firſt Hatched, it is like a little *Snail,* moiſt, Glutinous and ſhining ; then it waxes by little and little, untill it caſts it ſkin, and then appears its colour, and after that it again and again caſts its ſkin.

It changed the 22 d. of *September,* in leaves Knit together ; and the 12 *th.* of *July,* the Year following, came forth a *Butterfly Figured* in the *Table,* ; it lived, but two Days, though it had laid 9 Months, and 20 Days in the change.

Number. 93.

Tꝩe *Catterpillar* of the 83 d. *Table,* eats the leaves of *Labruſca,* (perhaps Bryonie) although with the ſmoak of that Herb *Mice* are ſaid to be killed: They eat not, as I have experienced of the Herb, before its Berryes come, and then they feed on the leaves untill the Berries grow Red and fall off; and then the leaves ſeem no longer fit to nouriſh them,

This *Catterpillar,* changed the 24 th. of *September* ; it continued in its change untill the 13 th. of *June,* the Year following, becauſe *Winter* came on ; and Flowers out of which *Butterflys* take their food, began to faile : the 13 th. of *June,* I ſay came forth a *Butterfly,* a of Yellowiſh colour, Figured in the *Table.*

Here the reaſon of the Noneruption *of the* Butterfly *all winter is, well aſſigned, to be the want of Food, and the aproaching cold.*   Num-

Number. 94.

The *Catterpillar*, of the 94*th*. *Table*, We fed with *Tobacco leaves* only; and it loves those leaves, which are largeſt and dry; and the ſmaller moiſt, and tender leaves, in the Wings of the greater, he cares not for.

He having cleanſed his body, changed the 3*d*. of *Auguſt* and the 17*th*. of the ſame Month came forth a beautifull *Butterfly*, which is wont to haunt Garden flowers; puts a long ſting into them, and is ſwift of flight.

This Butterfly *is frequent in* July, *to be Obſerved when* July *Flowers are in their prime; but is dificult to be taken by reaſon of its ſudden and ſwift flight.*

Number. 95.

The *Catterpillar* of the 95*th*. *Table*, in great numbers eat *Ozier leaves*, it is of an elagent colour, *viz.* Yellow diſtinguiſhed with Black lines, as is *Figured* in the *Table*.

It changed, the 8*th*. of *Auguſt*; and the Year following came forth a *Butterfly*, *Figured* in the *Table*. Of a Robuſt Body, and yet it lived but two days.

Of this alſo ſee my Notes.

Number. 96.

The *Catterpillar*, of the 95*th*. *Table*, delights in *Ozier leaves*, as many others do; to me the reaſon ſeems to be the dry temper of them; which corrects the moiſture

of

of *Catterpillars* If touched, it coyles up its body, like a
Ball; but you may put him out of that fearfull fit.

After it had well eaten, and well purged its body; it
changed the 10th. of *September*; and the 10th. of *June*,
the Year following, came forth a *White Butterfly*, marked
with Black Spots upon the wings.

Number. 97.

G. P. 1.
Tab. 63.

The *Catterpillar*, of the 97th. *Table*, eats *Rose leaves*,
and *Clary* : It changed the 20th. of *July*, as is *Figured* in
the *Table*; the 2d. of *August*, came forth a various co-
lored *Butterfly*, which I kept alive a while with Hony.

Number. 98.

F. P. 1.
Tab. 8.

The *Catterpillar* of the 98th. *Table*, eats divers Plants,
and are fonnd upon the Sand-hills; it changed the 12th.
of *July*, and the 5th. of *September*, came forth a *Butterfly*,
Figured in the *Table* : This *Butterfly* was tender and
weake; which I took to proceed from the *Catterpillars*,
being deprived too early of its food.

We found this Catterpillar, *on the Sandy Downs above*
Calice : *See my Notes.*

Number 99.

G. P. 1.
Tab. 17.

The *Catterpillar*, of the 99th. *Table* : Willingly feeds
on *Lettice* and *clary leaves*; it is very Hary, as soon as it
perceives any thing unusuall to it, it rowles it selfe up,
N   like

like a Ball, and moves not : It sets up its Bristles, like a *Hedghog* ; and if you take hold of its Haire, it easily suffers them to be pluckt off.

Sparrows, and other Birds eat not these *Catterpillars* ; but seem to abhor them, whence some guess they are poisonous.

It changed the 30*th.* of *June* ; and the 22*d.* of *July*, a large and beautifull *Butterfly* came forth marked with divers colours.

These *Butterflys* lay their Eggs before winter ; and because tthe *Catterpillars* are hatched late, about *October*, they therefore hide themselves in the Ground, chinks of Walls, and in other places, and live without food, untill the beginning of Summer, as I have had experience.

G. P. 2.
Tab. 23.

Number. 100.

The *Catterpillar*, of the 100. *Table*, eats the leaves of *Earth Nuts* ; with whith I kept him long : After his belly was fallen (perhaps for digestion sake) he ever crept long about the Glasse, I kept him under.

It changed the 3*d* of *August*, as is *Figured* in the *Table*, the 2*d.* of *June*, the Year following, came forth a *Butterfly*, to which I never saw the like ; Its wings were as White as Snow, its Body Yellow, and bright ; it flew swiftly ; first clensing its body of a certain humour ; it dyed after 5 Days keeping under a Glass, where it seemed to have been stifled, for want of a free Aire.

G. P. 2.
Tab. 26.

Number. 101

The Hairy *Catterpillar*, of the 101*st.* *Table*, eats willingly the leaves of *Heliotropium*, as its usuall food : They fight amongst themselves desperatly, and make the
Hiare

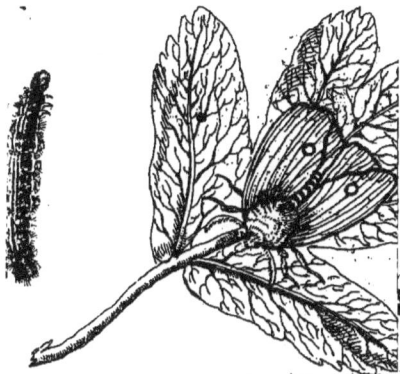

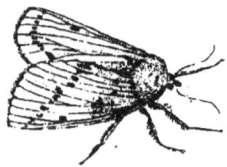

93.

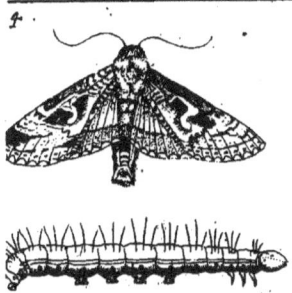

4.

95

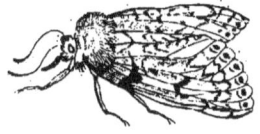

97.

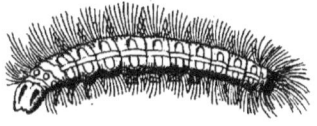

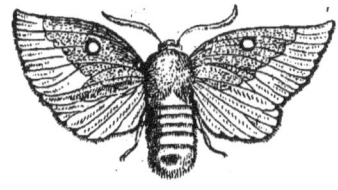

8.

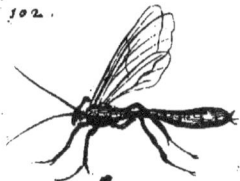

102.

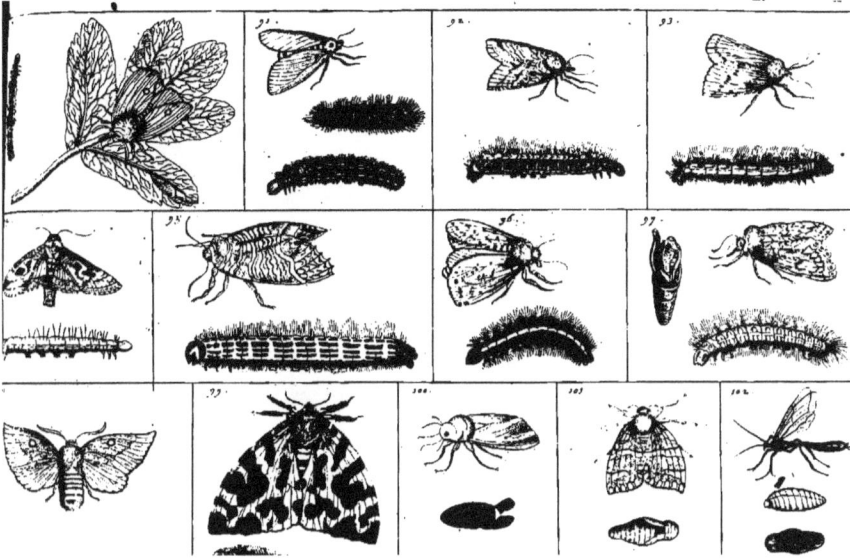

Haire Fly: They are very fearfull, and rowl themſelves
up when touched, ſleeping *Doggs-ſleep.*

It changed as is *Figured* in the *Table*: The 10*th.* of
September, in a bag of leaves knit together; the 10*th.* of
June the year following, came forth a *Butterfly*; which
ſeemed to ſleep ith Day, but was very ſwift and nimble
ith night, Bat like: I let it go, not knowing, how to keep it.

Numbar. 102.

G. P. 3.
Tab. N.

The *Catterpillar* of the 102*d.Table*, having had its fill of
Ozier leaves, betook it ſelf to a corner for change, as I
thought: But in ſtead of that I Obſerved a *Green Worm*,
to break forth of each ſide its body; this was done with
great anxiety, and contortion of its Body: The firſt
Worm, as ſoon as borne, faſtned upon the wound, it came
out off, and ſuckt up all the juice, and ſubſtance of the
Mother *Catterpillar*; ſo that nothing but ſkin was left:
The other Worm through the firſts greedineſs, being deſti-
tute of all food, dyed in the wound, out of which it
was coming: Whoſe body alſo, the Mother being evacu-
ated, was conſumed and drawn dry by the firſt Worm.
This firſt Worm being now ſatisfied, with the Bodies of
Mother and Brother, reſted moveleſs, and without any
other food, untill the 15*th.* of *October*, 1664. And then
changed into a Black Egg: The 18*th.* Day of *May*, the
Year following, 1665. Came forth two Flys, One of
of which is *Figured*: in the *Table*, in which *Hiſtory* it is
worth nothing: That of one Egg came two Flyes; that
by this meanes, the loſſe of two Worms, might be reſtor-
ed: Which as I ſaid, did make up the body of one.

Thus we have done with all the Hiſtorys *of* Butterflys,
I find in all the induſtrious Labours of Johannis Goedartius: *I ſhould not queſtion the truth of this laſt* Hiſto-

ry,

ry, *nor of the* Faithfullneſs *of the* Tranſlator. Mey *being e-*
very where as modeſt, *as the other* is *told and impertinent :*
I am willing to confeſs *that this* is *the only* puzzel *which*
I have yet met with in all the experiments of the Author ;
yet it may be, that two Ichneumon Wormes, *might club*
for a common Foliculus; *which he calls a Black Egg, as*
well as five, as in the 20th. Hiſtory, *aboue to be Noted.*
and yet that is, but a bare conjecture of mine, and which
this Hiſtory (*to ſay the truth*) dos not countenanc; *for it*
ſayes, that one Wormie *only, was remaining alive, which*
changed, into this black Egg.

Thus alſo *you ſee the* Butterfly *kind, takes up near two*
thirds of the Book; *but yet this was not the* Authors *choice,*
but his chance : which will appear, to him who ſhall conſider
that this kind of Inſect, *is for the moſt part laid,* Hatched,
and fed openly, and obviouſly upon Herbs, *and* Trees : *where-*
as the neſting, education, and food of moſt other kinds of
Inſects, *whilſt in* Embrio, *and in their ſeverall diſguiſes,*
is in darke and ſecret corners of the Rocks, *or within*
Trees, *and* Plants, *or under ground, or at the bottom of*
Lakes *and* Rivers : *So that, unleſs diligently ſearched for,*
they are not eaſily to be found.

Section. 4.

Of Butterflys *with* Tranſparent wings, *known by the*
name of Draggon Flyes, *in* Engliſh.

There is another kind of *Inſect, which the* Ancients *have*
left namelets : the more modern Authors, *have called them*
Libellæ, Perlæ, &c. *Theſe* Inſects *I reckon among the*
Butterflys, *at leaſt for ſome affinity betwixt them, I put*
them

them *nex after them: they all have* Foure Wings, *ſtiſſe,
brittle, and* Tranſparent ; *they are of a fierce nature, and
pray upon* Flys, *which they take in ſlight, as* Hawks *do the
little* Birds : *Some perhaps may think better to put them
after, or next the* Beetles ; *becauſe the* Worms *of both are*
Hexapod : *our Order will e agree well enough with both ;
and it is ſufficient that we do not confound them ; as being a peculiar* Tride *of* Iuſects *of their own kind.*

Number. 103.

G. P. 3.
Tab. R.

The *Hexapode Worms,* of the 103*d. Table,* when firſt
Hatched are very ſmall, like Gnats ; but in time grow
great : They are found in Ditches, and have no other
food, then that the greater feed upon the leſſer ; in a very
ſhort time I ſaw one of the three Large ones *Figured* in
the *Table,* devower ſix Leſſer ones of the ſame *Species* :
Hence it is, that theſe creatures are wonderfull timerous,
for if a Leſſer meet a Greater they fly a way with all
the ſpeed poſſible, but in vaine.

I find no more amongſt the Authors Papers, *but that of
one of the* Hexapode worms *Figured in the* Table, *a Sky
colour Winged Inſects;* Figured *alſo in the* Table, *was produced ; but I find not any one circumſtance of the mattrr
of the Change.*

Number. 104.

G. P. 2.
Tab. 14.

The *Hexipode worm* of the 104*th. Table* would not eat
neither Bread nor Meal, nor Sugar, nor divers Herbs
which I ſet before it : Nor would drink water : I then
bethought my ſelf of giving it to eat dead *Piſmierss*

N 3 *Beetls*

Beetles, and other dead *Insects*; I therefore put it up with a dead *Beetle*, into an *Ivory Box*, having learnt by experience, that these kind of *Insects* will perforate Wooden boxes; this *Beetle* it fedd on eating out all its bowels, and creeping into the shell of the *Beetle*; wherein it wou'd conveniently turn it self about every way: This Worme is slow footed : its an angry creature, and bold; keeping its station, unlesse you vex it overmuch : It hath a painted Taile, which is shining, with which it defends it self, that it cannot give back, and then it contracts its body, and with its toothed *Forcipes*, it strongly defends it self, against any violence done to it.

This *Worm* lay in the body of the *Beetle*, it had fed on, from the 18*th.* of *August*, untill the 8*th.* of *June*, the Year following ; then out came a winged *Insect*, beautifull, and elegant. The wings were very remarkable, but so tender, that I could scarce draw them with a Pencill, and shining bright as Mother of *Pearl*. Its two eyes glisterd like *Gold*; its body was of a Sea-green colour.

He flew with his head erected, and the rest of his body hanging downwards, with his wings spread, but not agitated; which perhaps he cou'd scarce do, as I guess from the bigness of them, but was carried by the Wings through the Aire, he lived but two Days fasting, for I knew not what to feed it with.

In

In the 5th. SECTION, *we put the* Bee-kind, *which have (* as all *the former* Infects *) Four* Naked Wing.

Section. 5. of Bees

Number. 105.

You have one of the great fort of *Bees,* (called *Bombilii,* by the *Greek*) Figured in the 106th. *Table.*

This *Bee* choofes moftly a deep and dry foyle ; to houfe in ; not a ftiff Clay, which wou'd not eafily be wrought into ; nor one two light, and Sandy, which wou'd be apt to fall in : alfo a deep foile they love, that the water may not foak thorough, and trouble them.

In fuch an earth they dig holes, to breed in ; fometimes three, and fometimes three foot and a halfe deep, and a foot fquare within : into thefe holes they carry fine, and dry Grafs, and cover that over with Wax,, no otherwife then *Swallows coat* their *Nefts* with Clay ; but thefe Bees make their Nefts *Tranfparent,* and full of holes ; and do coat them over, very neatly, with coverings made of Wax ; that they may be fafe from Rain ; which might otherwife foake through the Earth and get into them.

In this Neft they make a *Theca,* or fmall *Cell,* like the halfe of an Egg, and the bignes of a Pea, divided in 2. parts.

Every Bee lays 9. little *Worms* in this *Theca,* or *Cell,* of the bignefs of a Muftard feed, but as white as milke, and *Pellucid.* Then they fhut up this *Theca,* that it may
be

be like a compleat Egg : This *Cell* now to the Senſe is moderatly hot, as an Egg ſat on by a Hen.

I Obſerved the leſſer Bees often to creep over, and about this *Theca*, or *Cell*; as though they had a mind to keep it warme, and ſo hatch it; and again I ſaw them ſcratch it, as though they wou'd make a crack or opening in it; but they made it ſofter by ſcratching it.

The *Worms* included in this Egg-like *Cell*, daily grew biger and biger; in the mean time the leſſer Bees did not ceaſe their indeavours to hatch it, nor did they leave it off, untill it was broke aſunder; alſo the *Worms* within by the continuall agitation and motion of themſelves, did help forwards, this work of breaking it open; and they growing; the heat alſo increaſed, which made the Wax yet more ſoft.

The *Worms* hatched from this Egg or *Theca* lay quiet a whole Day, and then the great or *Old Bee* coming, devowered all the Wax, of which the *Theca* was made, and in which theſe *Worms* were included; that this Wax being again liquified in her body, might ſerve, as the materiall, of another *Theca*.

Now when theſe *Worms* (thus hatched out of the firſt *Theca*) have layn a whole Day, quiet without moving at all, and like dead things, they weave each for himſelf an Egg, of the bigneſs of a *Bay-berry*, of a certain ſpittle, deſtilling from their mouth, and body.

And being now made, they are indeed, moiſt and ſoft; but in time they dry and become hard.

Theſe Eggs are all glued together in a *Bunch* by theſe *Worms*· and out of each Egg comes a Bee, as is *Figured* in the *Table*; where the Eggs or *Foliculi* is opened at one end, that the Bee may be ſeen, which comes of the *Worme*, *Figured* alſo in the *Table*.

This done the Great Bee comes again, and makes a new *Egg*; and puts therein 5 *Worms* of her own laying, and then another Bee do's the like, and 3*d Bee*, and ſo the reſt

follow,

follow, doing in like manner; untill all the work be finished
and made like a bunch of *Grapes*, in forme and Figure.

And yet whether one and the same *Bee* perfect this work,
and lay *Wormes* in all the *Theca's*, or Cells of the *Bunch*;
or whether they are divers, and many *Bees*, at this work,
making, and filling the *Theca's*, with *Worms*, we are not
certain; but we judg it likely, that one and the same *Bee* re-
quires the making of the whole bunch of *Theca's*; and fills
them with her *Worms* only, for with these Eyes we have
seen one *Bee*, which alone layed 33. *Worms*, and put them
into the *Theca's*. This is also Observabe, every *Bee* in the
making of every Egg, is busied, before that be finished, for
28 hours in a continued labour almost, yet some get their
work done sooner, others later a little, as it is with Work
men, which are more handy some then others.

The *Worm Figured* in the *Table, changed,* the 11th. of
July, and so continued, untill the 17th. of *August,* when
it appeared in forme of the *Bee, Figured* in the *Table.*
when these *Young Bees* have, by biteing, opened the *Wax
cases,* and are come forth of them, they are wont to rub
their Eyes, with their foremost feet, and by this means of
rubing, they seem to open their Eyes, or at least to rouse
themselves from sleep; in the fore part of their body,
they are *White, Te lowish* also, and *Black,* their wings ly
upon their backs, and are not yet expanded, and they are
moist; but are soon dry, by the Air; that n a quarter
of an hour, they can expand, and extend them.

These little *Bees,* new hatched, are not yet fit for worke;
but yet they covet to get up into the open air; but are a
hinderance to the greater *Bees,* which are busily employ-
ed, who therefore drive them down, as often as they
come up; also they run too and fro, as though they were
drunk, after three Days that they have been hatched,
then they are able to carry Earth, to the nests, which the
old *Bees Arch over,* with Earth heaped together, which
Hollow Arch, they draw over with a coat of *Wax,* as

O

Brick

Brick-layers Servants carry Morter, and Bricks with which they build Walls: In like manner, these *Young Bees* serve the Old ones; but they do not carry Earth forwards; but like *Hens* scratch it backwards, and those *Young Bees* which were first hatched and are elder then the rest, are imployed about the more master-like works; for they make Wax, dig holes, and Arch them; the Younger geting Earth; The elder eat Wax and soften it in their stomacks, and throw it up againe, by little and little Liquified; and of this *Liquid Wax* and Earth mixed together, they make *Nests*; not unlike *Swallows*, and finish their work with their fore Feet complicated, and the Younger help them in bringing and ramming the Earth.

This *commonality* also has its *Kings* or *Captaine* as the Noble *Bees*, by whom they are governed, and this *Bombylius* or *Master Bee* is very old, and in body *Greater*, then the rest, like as the *King* of the little *Bees* do's excell the rest in magnitude, and shew: But this is destitute of Wings and Hair, where the rest abound, with *White*, *yellow*, and *Black* Hair, and are rough; this is alltogether Bald, like naked Birds, or as is painted on the back part of the Head; moreover he is Black and shining, as Polisht *Ebony*; sometimes he coms to look upon the Works in which the *Commonality* is busied, and to explore whether they are made of a just measure, as well *Latitude*, as *Altitude*, conveniently creeping into the Nest, and creeping out; Ascending, and Descending, he do's seem, to measure as *Architects* are wont to take an exact account of the work, which the Work-men make: the *lesse Bees* when they meet this *Old Bee*, do not exhibit a little honour to him, as to the *King*; they do run about him every way; they stroke him once and again; in the mean time playing with their wings expanded, embraceing him with the *Anteriour* feet, as with armes, even as we gratulate him we meet, therefore

the

the *Bees* are an example to us, of giving honour to old men.

When this *Captain Bombylius* has explored whether the work answeres the Rule; he doth descend into the *Latibulum*, and every one of the younger does anew set upon each his work.

In the Morning these come late to the work, and unless these little creatures, which daily in the morning creep upon them and about them, did not excite them, they would come later.

They have also amongst them another *Bombilius*, who like the *Drummer*, do's beat to the Souldiers, to come to the Gardes, or to move the Camp, or to March, or to Fight with the sound of the Drum, so also do's he excite them to work.

This about 7 a Clock ith morning, do's ascend to the top of the work, to be built, and with the halfe of his body, looking out of the hole, doth vibrate and agitate his wings, and by the benefit of them, doth excite a streperous noise in the hollow *Latibulum*, not unlike the noise of a Drum, which mostly is heard a whole quarter of an hour.

I, an expert person, and an Ear witnesse, do commemorate these things; Yea and an Eye witnesse, and together with Me, many others have heard it more then once, the Vibrating and Streperous wings of that Drummer of the *Bombilii*: I say, Persons, which for the hearing of this curiosity have visited me.

Besides, and to these, there is one that doth *Watch*, for when sometimes I did beat the nest, this Forth with did ascend out of the *Latibulum*, as one struck and astonished, and did *Discurre* every where, as one that did explore what matter was without: but suboderating noe danger, again he soon descended. More then once I have with admiration seen that. Sometimes also, I have put upon the stick, a Domestick *Bee*; but having spyed this,

forth-

forthwith out he came, and fnatcht the *Bee* as one in
wroth, and left it not, till it was killed.

Furthermore I have found thefe *Bombilii*, to coat
with Wax the Cells, out of which their fellow *Bees* had
gone, to fill them with honey; to feal them up with
Wax, left the Thieves entering into them, fhould waft
the Honey. For amongft thefe are to be found idle
Drones, Thieves and flow *Bombilii* : Yet we have feen
thofe, with the reft going abroad to colleſt Honey, did
make a fhew to fly out with them, but they did not goe
out of the Chamber, in which I did keep them ; but
did only fly a few rounds, and that being done they did
returne into one of the Glafs bottles : (For I had placed
two in the Chamber) in the which they had built their
Combs : And each one returned into the Bottle out of
which they came forth ; when the reft flew out into the
Fields, and Meddows, and afterwards did returne home,
fome loden with Honey, others with Wax, and fome
with Water, thefe Droans did returne empty into the
Hives :: and moreover did devour the Combs made by
others, and the Honey gathered for winter.

Thefe are the idle Cattle which do not imploy themfelves in the gathering of Honey ; but do confume the
aliments gotten together, by their fellows, concerning
which the *Ancients* have faid much, *&c.*

Perhaps you will afke what was the end of this Old
Bombylius, which I did think to be the *King* or *Captain*:
Him I faw near dying early ith morning before the
Drummer did Vibrate, as he was wont, his wings, and
did excite a noife to them, creeping out of the Houfe:
But wholy horrid and Trembling in body ; hither come,
as it were tired, he lay down and without Convultion
of the feet, fweetly did expire ; neither did that feem a
wonder to me, for his feet were long fince numbed with
Old age.

The *King* being dead, the number of the other *Bomby-*
lii.

lii, did daily decreafe, at laft I did fee a *Bombylius* creeping out of the Cmbs, whofe Head was pluckt from the fhoulders : A body without a name.

I beheld this Trunk palpitating, and lying alive two Days, and then expiring, hence it was eafie to guefle, all things within to be troubled and done inordinatly and ill, and the *Bombylii* to doe as they lift, and to fight fiercely the *King* being dead ; for from that time the *Drummer* beat not any more, as he was wont, the *King* being alive.

It is alfo worth admiration , about the birth of thefe *Bombylii* from one Old *Bee* of this kind, when he do's compofe himfelf to propagate his *Specis* , Sometimes to come a *Hundred* and *Twenty Worms,* yet fometimes fewer, and fometimes more ; but one only fometimes, to have bred at once , there are many Eye witnefles befideds my felf, who have wondered, one of thefe little *Bees* to be fo fruitfull, as to be able to conftitute a *Bee King-dom,* and an *intire fwarme.*

Section. 6. of Beetles.

Number. 106.

G. P. 2.
Tab. 54.

Betwixt the bark and the body of an *Oak,* growing in the Ifland *Wallacria* in our *Zeland,* I found this *Worm* boaring it ; though foft, knew how by the benefit of a peculiar organ, as with a wimble to perforate a very hard wood ; and this kind of Worms ufe their mouth, out of which a fharp and hard Inftrument doth ftick for a *Wimble.* Alfo they are not a little ftrong, in protracting the hinder part of the body to the head wards : Moreover

O 3   they

they do affix the *Podex* in the excrements, and so they are
placed betwixt the Wood and excrements, as it were
squeefed in a Prefs; fo that they may force with their
mouth and boar, and in this pofture they do pretorate the
Oke, and do live of the marrow and fattifh matter, which
they boare out. Being fed to fatiety, they do extend
themfelves ftraight, when the Ventricle has done its
office, they do exonorate the belly, and at the fame time
they do, by little and little, more and more contraft the
body, and draw it up, as much as may be; fo that always the voided excrement is found very clofe to the
body : And fo preffed they force into the wood.

This did give it felf to *Tranfmutation*, the 28*th.* of
October, placed upon the bark of the *Oak*, on which
I had found it fitting, together with many other little
ones; yet I judging it alive (for I faw them wag frequently) the Year following the 13*th.* of *January*, it
did produce an odd *Animall*, of a Hoairy colour, variegated with black fpots; but yet ftout and fierce; in the
Forehead, it was Armed with Horns, which it did draw
back, when it was angry or did move forwards; but did
joyn them, when pleafed; did not fuffer it felf to be in it
clofed in a wooden box, for it did throughly bite into the
wood, and that quickly, and fo broke out, and fo did
fhew from what kind it fprang; for I cou'd not explore
of what aliment it did live, and therefore cou'd not keep
it alive, but I fuffered it to ftarve.

E. P. I.
Tab. 66.

Number. 107. *b.*

The *Worms* of *Table.* 107. *b.* are moft bitter enemies
of *Caiterpillars*: The fore part of the body has two open
claws; whichfhut and open like Pincers; by thefe claws
they pinch the bellies of *Catterpillars*, and hang upon
them

them ; the *Catterpillars* feeling the pain of the wound, do move and agitate the body every way, that they may free themselves : In the mean time the *Worm*, like one dead, the body being extended, doth quiesce without motion, and by how much more the *Catterpillar* doth move, and tofs ; by fo much more is it hurt, and the belly is burſt ; after that the *Worm* has let go, the wound doth forthwith ſwell, which ſeems to indicate a venome.

This *Worm*, well armed by nature, is Yellow ; and of a ſplendid colour : it cannot eaſily be hurt of the *Catterpillars* ; it cannot live above ground above two Days.

I put upon the ground one of the *Worms* deſcribed almoſt dead ; which forthwith recovering ſpirit, penetrated by boaring the ground.

In winter time diging, I found one of theſe *Worms*, the Ground being Frozen, aboue two foot deep, (they eaſily abide cold,) together with a certain Bee, which I placed near the *Worm*, that I might obſerve how they agreed.

The Worm, forthwith ſet upon the *Bee*, and took hold of his head with his Pincers, and ſo long agitated, untill the *Bee* very much tired, did endeavour to get away ; but his wings being frozen, could not.

A little battle I ſaw at the ſame time betwixt a *Bee* and a *Catterpillar* ; which I had then alſo dug out of the Ground.

Number. 107. *a*.

G. P. 2. Tab. 19.

The two *Worms* expreſt in the *Table*, are plainly of the ſame nature, and caſt upon the fire, do conflagrate like *Oyle*.

Now I ſhall relate the Metamorphoſis of them.

One

One of the Worms, when he lay ftill for Transformation, it was the 2d. of *June*, and he continued in that
pofture till the 24th. of *September*, in which pofture he is
exprefled in the *Table*.

But that fame day he brouhgt forth an *Animall*, like to
a *Chryfalis* : as that Worm had a pair of Pincers, fo this
Animall was in like manner armed with Pincers, very pernicious, for with them he pierces and breakes the *Eggs*
of *Ants*, and *Gryllotalpæ* ; wherever he finds them : He is
equally as malicious, as the Worm, from which he is *changed* : Alfo he fiercely fights with his like ; and efpecially
after three Days hunger ; for then he fights moft ftoutly
with his Princes, *&c.*

When you do caft him *Ants* or *Gryllotalpæ's* Eggs forthwith he doth *Exuge* them boared through with his beake.

As this Animall is an enemy, to all *Infects*, and do's
endeavour to devour them; So alfo it ha's its enemies
and indeed very *Infeft*; to wit, the *Gryllotalpæ*, for
thefe every where near their neft, make covered Bur-
rows in which they do wait in Ambufh, and obferve
which they may devour. As thefe are wont to kill *Catterpillars*, fo to them is returned, what they have given
to others : Yea as thefe have treated the *Gryllotalpæ*, not
yet excluded and hidden in the Eggs ; fo alfo the adult
Gryllotalpæ, do treat thefe, and prevaile over them.

Thefe *Worms* are two years old, when they *change* ;
neither do they *change* the *Huckle* : But when they *change*
forme, then *Snake like* they put off the *Skin*, and grow
Whitifh ; and alfo on each fide the beginning of the
wings is then beheld to come forth, but the Head and
Feet obtected with a certain thin membrane, or fkin,
which by little and little doth excrefce with the Feet,
and at length doth fall off.

And this membrane doth feem only given to them for
the confervation of their members, when they are in the
ftate of Transformation ; and now when thefe *Worms*
are

are plainly Tranfmuted as dead, yea void of all motion,
they ly upon the ground fixed, and there appears no figne
of life in them.

Number. 108.

G. P. 1.
Tab. 6.

The *Worm* reprefented in *Table* 108. Was bred from
the fweet Root called *Skirrets*.

In the fame Root in which it was bred, it compofed it
felf for *change* the 9th. of *Auguft*, and the 25th. of the
fame month came forth a *Black* creature depicted in the
Table : it is of a flow gate, as the Worm from which it was
bred, when this little beaft wasfirft feen, it was covered
with a Yellowifh colour very pale ; the *Fore-part* was
Reddifh, but afterwards it became exactly Yellow, and
then of an *Amber* colour; and at length elegantly Black.

I kept him alive fome Days with fweet juices.

Number. 109.

G. P. 3.
Tab. M.

This *Species* of *Teredo* is wont to be found while yet it
is little, betwixt the Bark, and the Okewood, or alfo in
the very Bark ; and there it is procreated of Seed ; but
now having got more ftrength, it do's boar into entire
Trees; by its moft fharp mouth, and doth inferre great
damage to them : it doth feem a wonder, or at leaft worth
the Obfervation, a *worm* fo foft, little, and tender, to
penetrate fo great and moft hard trees, and boaring the
trees do's follow after the fat, and oleaginous juice with
which it is Nourifhed.

It is neceffary that the Anteriour parts of the Mouth,
to be not only hard, but alfo fubtle, and fharp; by
which they may penetrate to the intimate parts of the

P

Trees

Trees, and to *comminute* their substance, although wood, and most hard, into a most fine Powder; which in every *Species* of *Teredo* may be observed.

It composed it self to *change* the 10th. of *November*, 1663. And the 6th. of *Aprill*, the Year following 1664. the Red creature presented it self to our Eyes designed in this *Table*.

I kept it alive with Sugar untill the 5th. of *May*.

G. P. I.
Tab. 79.

Number. 110.

The *Worm*, of the 110th. *Table*, was brought in a certaine Ship from *New Zeland*, in the *West Indies*, into this Country.

I found it in the bark of wood, of which Chests are wont to be made, in which is wont to be brought *Sugar*, out of the *Indies*; from that bark it hath its *Originall*, and doth seek nourishment; it began to undergoe the *change* the 7th. of *September*, and remained in it to the 19th. of *October*, and at that time *changed* the skin; and with the skin the shape; by little and little, the colour, members and all the rest.

The manner of the *change* (as I Observed it) I have set forth in the *Table*, and at length, the Transformation being compleated, an *Insect*, (perhaps never seen in these *Regions*) came forth, of a wonderfull structure and shape, much unlike to the first which it had; which I have also delineated exactly.

G. P. I.
Tab. 78.

Number. 111.

The *Worm*, of the 111th. *Table*, is in *Dutch* called the *Corne Worme*; because it consumes the roots of *Corne*; it is found also in Gardens and Orchards.

I

I took this *Worm* the 22*d*. of *August*, 1659. And I kept it a whole Year in a Glaſſe bottle, with Earth put in the bottome : To which I injeᶜted the ſeed of *Henbite*, with a *White flower.* (for there is another Herb of that name with *Purple flowers*) and I obſerved in the evening the *Worm* to be wont to come forth, up from the bottom of the Bottle, that it might feed upon the Herb, and Flower, before deſcribed ; this being done, again it hid it ſelf within the Earth ; for never or very ſeldom, it do's appear, in the Day time, above ground.

After the *Wormes* of this kind have eaten enough, and have come to the juſt Magnitude of body, they ſeek high places ; that they may ſafely quieſce, and that they may pleaſingly compoſe themſelves to Transformation, which they expeᶜt.

This *Worm* ſeems to come from the ſeed of thoſe *Beetles*, which eat the leaves of trees, and which is very frequent in *Holland* upon the tops of Trees in *May*, for then their nouriſhment every where abounds.

This kind is wont to ſtay above ground two whole Months, or a little more, the reſt of the Months of the Year, it lyes hid under ground and uſeth no nouriſhment ; like to a dead thing, and it do's always, unmoved, plainly quieſce ; but touched with a hot hand, or otherwiſe, it forthwith ſtirrs ; as I have often tryed. And this alſo of this kind of *Beetle*, never to be found two or more joyned under ground ; but alwayes *Solitary*.

Before the *Worms* hitherto deſcribed, were transformed into *Beetles*, they had lived over the forth Year.

The *Worm* of this *Table*, did begin to *change* it's forme on the 3*d*. of *September*, 1658. and I have deſcribed the manner of it's Transmutation ; and in *May*, 1659. from that came a *Beetle* depiᶜted alſo.

It lives long, provided it be not ſtarved ; for want of food ; or through the vehemence of the cold.

Number

Number. 112.

G. P. 2.
Tab. 18.

This Kind of *Worm*, I found creeping upon *Green-corne*; and in vaine have fought it other where : They are pernicious creatures, for they eat the green Ears or Sheathings of Wheat : They fwiftly afcend, and defcend, as foon as they perceive the Corn to be touched with a Stick or the Hand, (The like is above noted of an *Eruca*) forthwith they caft themfelves upon the ground, be they little or greater; the leffe fwiftlier, then the greater; for that thefe are not fo Tenacious of the Corne as thofe, which are Armed in the hinder parts with *Clikers,* by the help of which, they hold hard upon the Corne; being fallen upon the ground they creep into it, and ly hid under it.

One of thefe, affixed to a ftalke of wheat, being about to *change* into a new forme, the 22*d.* of *July*, as is depicted. On the 8*th.* of *Auguft*, came forth an *Animall* very thirfty, for as foon, as it came forth, it drunk much, and often.

I learnt by experience, (and all that I have writ I have proved fo) this Animall may very long be kept alive with water; and Sugar; but deftitute of water; it lives not above four or five Days.

C. P. 2.
Tab. 15.

Number. 113.

Thefe *Animalls* for the Elegancy of them, are called *Lady-clocks* ; they owe their Originall to the Seed which their Parents fhed and put upon the *Curran-bufh leaves* ; and which the heat of the Sun hatches.

Thefe *Eggs* like feeds layd clofe in a round, as loaves in an *Oven*, the 20 of *May*, firft did look black: But the
29*th.*

29*th.* of *May*, did grow greenish at the bottome; and the Second of *June*, every one of these Eggs, had a black spot, or point in the middle; and the 5*th.* of the same month, Animalls were hatched out of these *Eggs*; partly yellow-ish, and partly black; in which no touch of life did ap-pear, unlesse breathed on, then they moved, or rolled, to and fro.

But the 6*th.* of *June*, when now their feet are grown, and hardened with the benefit of the aire, they creep a-bout; then run about; but about evening, and when the aire was cooler, they run together, and kept close; either for company, or for heating one another; and in this situation, and posture, they remained till 6 ith Morning of the day following; then every one set out to seek food, and that, the dew which useth to fall upon the leaves from the skie, and endued with a peculiar force of Aliment, well knowne to those, and other In-sects, for food; for they feed on it, &c. About the 13*th.* of *June*, these *Insects* did put off their skins, Ser-pent like; which being done, they are Yellowish and Blackish.

The Day after, *viz.* The 14*th.* or about it, they *change* forme another time, and colour, so as to grow Yellow and Reddish.

They begin the 3*d. change*, two Dayes after the Se-cond *change*, in that very posture, and with a Black and Red colour.

The 4*th. Transmutation* they are wont to undergo, the 3*d.* Day after the 3*d. Metamorphosis.*

And this is the last *change* which they suffer, and by it they get their last skin, partly Black and Red, and where it is Red, variegated with Black spots, or points, as is depicted.

Tis worth the noting, as often at they cast their skin, they fix their feet fast to the Leafe or Paper, they chance to be on, and when they have fixt them, they creep out

of

of the skin, and leave it standing so, that seeing of it, you
woul'd say it was verily one of those Animalls standing,
&c.

I suftained these Creatures for some time with dew,
collected before Sun-rise from the *Curran-bush leaves*, untill I saw them to have attained to their full Magnitude,
and perfection, nor to make any further *change*, and then
I set them at liberty to provide for themselves.

G. P. 2.
Tab. 41

Number. 114.

With this fort of *Worm, Skins* and all things made of
skins, are much infestd.

This I found upon the Feathers of the dryed body of
a *Duck*, eating the fat and skin thereof, and perhaps they
are bred of it, because they feed of it, as many *Insects* e-
xist from that, with which they are nourished, as the *Philo-
fophers* Write. They creep very swiftly, as soon as they perceive any thing, which may trouble them; they hide themfelves forthwith under the Feathers, so that they may not
easily be found.

This deserves Admiration, these Worms to cast 10
Skins, before they *change*; and as oft to cleanse, as they
cast their skins.

Their Excrement is like a slender Thred twisted, yet
firmly compacted, a Span long, and which may be extended, for it is *Vifcous*, and hard to break, and under dirt,
they are wont to ly hid.

I tryed this Worm to have fed, from the 10th. of *June*,
to the first of *September*, when it did desist from eating, it
began its Transmutation, as is depicted *Number* 114. and
the 20th. of *October*, to have bred first a Hair coloured
Animall; which then put on a Red, and lastly a Black
colour. which was Yellowish, in the middle of the body.

You may keep these Anmalls, as long as you please if
you

you give them *Walnuts* to feed on; but take their food from them, and they live not above four days.

No Infects are so fearfull as these are, for let them have the least perceivance of any thing, that may hurt them, and they presently counterfeit themselves dead for fear of being killed; and indeed, it is so with all kind of *Animalls*, who naturally fly things hurtfull.

Another *Worm* of this kinde I fed, not unlike in colour to the other, *Figured* in the same *Table*.

Which the 3 *d*. of *September*, composed it self for *change*, in the *Figure* depicted, it remained in that state to the 20*th*. of *November*, on which day it boar Twins a Male out of the right, and a Fe-male out of the left side.

But you'l ask, how knew you to distinguish *Male* and *Fe-male*? I answer, both these within a quarter of an houre after their birth copulated, and shewed me which was *Male*, and which *Fe-male*.

The *Male* was longer, and slenderer, then the *Fe male*; which was shorter and rounder, the one delighted in the others company exceedingly; but I let them go together at liberty into the Fields.

I was an eye witnesse of this secret of Nature, from one and the same *Species*, two specifically different *Animalls* to have been bred, from the one, One; and from the other two a *Male* and a *Fe-male*; I saw this Transmutation with my own Eyes, and coul'd not find the reason of it.

Number. 115.

I find by Accurate search, that about the midle of *June*, a Greasie and Fat substance doth drop out of, both the upper, and the under side of the leaves of *Moth-mullen*. Sometimes sooner, sometimes later; but the 15*th*. of *June*, I found out this thing.

This

This fat juice did become a living *Animall*, the 20th. of *June*, to wit, a little *Worme*; which was in the middle of the Body pellucid, like water; was of a *Saphire* colour, Greenish, which slowly and gently was agitated by the Suns heate; but the 24th. of *June*, it became of a Yellowish colour, and the 27th. of the *fame Month*; I did distinctly see on the hinder part of the Body, which was Yellowish, 32 Black spots, and forwards on the body 8 more.

These *Animalls* feed of the down onely growing on the leaves of *Mullen*; and this wool, or down mixed with *Venice Turpentine*, and used by way of suffumigation in smoke, cures the Piles; which I have tryed.

This little *Animall* did give it self to *change*, the 30th. of *June*, the same Year; but because nothing particularly came from it, I have not depicted the forme of the Sheath, or *Chryfalis*; for only after it had given it self to rest, it grew in its roundnesse, as though it would *change*, and the 8th. of *Auguft*, it did creep out of its skin, as out of its Shell, Yellowish distinguished with Black spots, and round in *Figure*.

It is worth observing, that these Creatures, when they first grow Yellow, ever before they have feet, are besieged by their enemies; and they are certain little *Spiders*; which are of the same colour with them, and bignefs; and without doubt, deceive them, upon the likenefs of them.

These *spiders* are bred of the same Leaf with them.

Moreover there is another Creature of a Black colour with Pincers in his Forehead, which he opens and shuts, as he lifts; with which he kills these *Worms*, and *Spiders*: This also I guefs is bred from the same Leaf, *&c*.

All of them feed of the Down growing upon *Mullen*. This Down is pellucid like *Chryftall*.

Number.

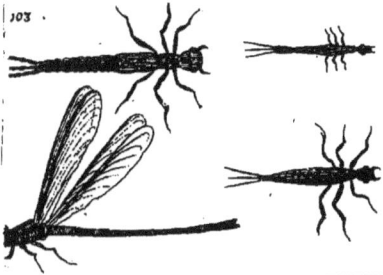

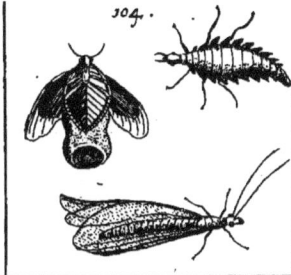

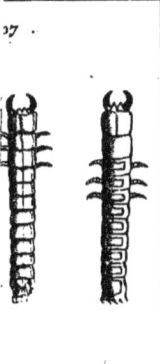

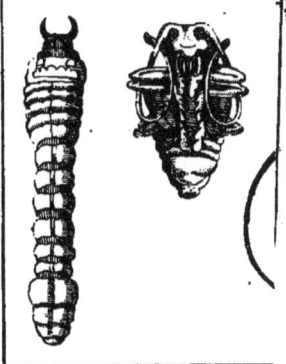

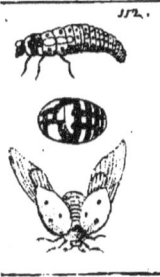

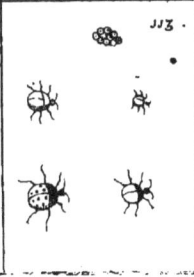

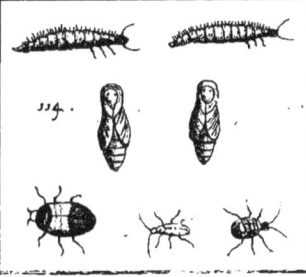

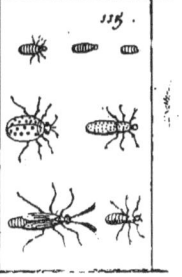

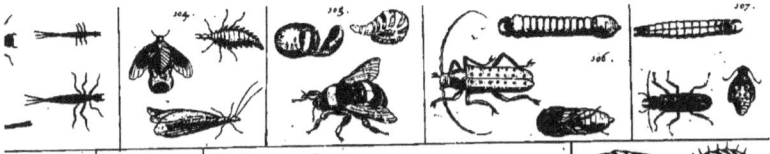

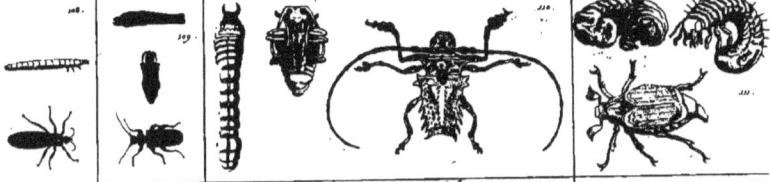

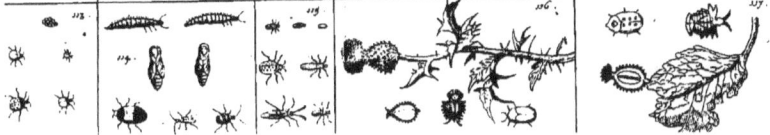

Number. 116.

G. P. ɪ.
Tab. 43.

This little *Annimall* feeds on a *Thiſtle.*

Hee began to be Transformed the 9th. of *July*: The manner of the *change* is depicted, for its ſingular elegancy; The upper part repreſents the *Figure* of an *Imperiall Crown*, under which the Image of a humane Face offers it ſelfe: In the middle you have the Image of ſome enſign.

The ſtate of that Transformation continued in this manner 12. Days; after that appeared a great *Animall*, having ſix Feet; which alſo I painted.

Number. 117.

G. P. ɪ.
Tab. 44.

This *Worm* is not much unlike the former; *Meliſſa* Yields it aliment; whoſe juice drunk with Wine, is believed, to mitigate the pain of the Stone.

In creeping appears, near thee xtream parts of the Body, a thin ſkin; which do's ſeem to refrigerate it by continualy fanning it.

It began to *change* the 7th. of *June*, and after 10 Days it received the forme of the *Inſect*, depicted.

Number. 118.

G. P. ɪ.
Tab. 45.

The *Worms* depicted in the 118 *Table*, I found upon the leaves of the *Willow*, in a certaine order; as it were, diſpoſed in *Battle Aray*, and with the bodies erected.

And when they had come to their juſt Magnitude, their bodies being inclined, they began to eat, and to

Q   abrade

abrade the tender particles of the leaves, foe that nothing was remaining, befides a certain dried fkin.

They appear Yellow in the beginning, afterwards
they become black.

They compofed themfelves for *change*, the 7th. of
June, and the 18th. day of the fame moneth appeared
the Animalls depicted.

Thefe *Animals* have a certain glew, near the Tail ;
by which they do adhere to all things fo Tenacioufly,
that they can fcarce be fhaked of.

G. P. |1.
Tab. 76.

Number. 119.

For *Gryllotalpæ* or *Field Crickets*, I know them by
many experiments to be very rebuft, and of a firm life.

I cut off the head of one of them ; which after 2 dayes
was wholy eaten by another *Field Cricket* : onely 2 little
Nerves being left, yet the head lived 12 hours after.

I hanged another *Field Cricket* by a ftring in the heat
of the Sun, fo that it became wholy Black : yet it died
not, before the 7th. day.

They are very ingenious in building their neft : for
that end, they do elect a certain *Glebe* of earth firm
and Tenacious : and therein they make themfelves a hole
to go in and out at ; within they make a great cavitie,
in the which they depofe more then 100. Sometime
150. eggs.

This being done : the hole of the ingreffe they accura-
tely fhut up, and they ftrengthen the *Glebe* as much, as
may be ; for that being broken, all the eggs perifh, and
are confumed of certain *Black Flyes* ; which ly hid under
ground ; therefore they are very anxious and folicitous
of conferving, and making firm this *Glebe* ; therefore
they perpare for themfelves a certain fubteraneous *Du-*
ctus,

Ctus, round about the *Glebe*, that they may go round a-
bout this *Glebe*, and preserve it : besides, about this *Glebe*
they have other holes, and hiding places ; whither in time
of necessity they may Fly.

Again, they know how to raise up their nests, by a
wonderfull industrie, in a hot and dry season, that they
may almost touch the superficies of the earth : that by
so much the better and sooner, the egs may be cherished
with the heat of the Sun, and hatch ; on the contrary,
the air enclining to cold and humiditie, they do sink
lower into ground their nests.

I have observed also the *Field crickets* to have wings,
but not to fly : but for ornament ; that with them they
may cover, and preserve the very tender hinder part
of the body.

In the *Island Wallachia* of *Zeland*, there are many
Field Crickets, and they do much hurt to the young and
tender corne, which they saw in two with their mouth
and cut the roots.

The *Gardiners*, that they may remove them, put into
the ground, little pots, that the upper lips may be equall
with the superficies of the ground, the *Field Crickets*
falling into these cannot get out.

Or their nests are to be broken, and the eggs spoiled

Number. 120. *a.*

G. P. 2.
Tab. 42.

The Animall depicted in the 120*th. Table*, *a.* is ugly.
Beetle-like, and mostly lyes hid underground ; it has no
wings, it feeds of the leaves of *Anemone*, or the garden
Ranunculus.

This I put into a Big Glasse, the 5*th.* of *May*, filled
with new earth ; the 11*th.* of the same moneth it thrust
it self, with the hinder part of the body into the earth ;

Q 2   at

at firſt I did not obſerve that hole was by it, but take-
ing it out, I made all plain again.

The next time I looked at it, I obſerved, that it had
thruſt its body again into the earth, and I took it out a-
gain ; but the 3d. time ſeeing it again in the hole, I let
it alone, & did obſerve not-a-few Eggs ; which it had
layed in the hole, Yellowiſh *Eggs*, like grains of Sand,
thick compacted, as in the *Table*, they are depicted. When
this Animall had layed its ſeed, it was as little again, as
it had been, juſt before ; for it cou'd ſcarce draw its load-
en and big body , and now it became more *Agill*, and
eat more meat, and greedilier, then before : alſo in a
night, the hinder part of the body was extreamly ſhrunk
up, and ſhortned by the halfe ; it ſlept at the leaſt 12
hours daily, as oft as it awaked, it devoured its uſuall
food, which I gave it, greedily.

And the 2d. of *June* , I found it again crept with the
hinder part of the body into the earth, continualy beat-
ing the earth therewith, that it might more eaſily lay
its ſeed *&c.*

After this manner it placed its ſeed upon the earth
the 2d. time, in the ſame form as the firſt ; but that
being done, it lived not long after ; for when the leaves
of *Anemony* did wither, it died with hunger, without
doubt.

From the ſeed of this Animall, which was firſt layed,
many very little *Wormes* were hatched, not thicker, then
horſe-hairs : But the 21th. of *June*, the crept ſwiftly, and
were as big as Muſtard-ſeed, had ſix feet and 2 horns.

The greateſt part of theſe *Worms* being numbred, I
found them above *Two Thouſand* : but I believe there
were 3 *Thouſand.*

The ſeed which it layed the Second time, the 29 of
June produced alſo many *Worms*, juſt like the former;
but not ſo many in number, to ſatisfie my curioſity, I
took them up on my Pencill point, a little wetted, one
by

by one, and put them into a glafs, and thus I counted
906. but many I loft in the telling.

I took great pains to find out a gratefull food for
thefe *Worms* ; but to little purpofe, I gave them honey,
dead earth *Worms*, Ants Eggs, Sugar, Bread, divers green
herbs; but I cou'd not bring them up to a juft bigneffe
with all thefe.

But by means of a *Microfcope*, I Painted to the life one
of them, in the form you fee it in the *Table*.

This and others I kept alive, and found that yearly
they *changed* colour, the 2*d*. year brownifh, the 3*d*.
year black.

And thus much for the female *Beetle* I fhall obferve
fomething of the Male of this kind.

Number. 120. *b*.

G. P. I.
Tab. 74.

I found the *Worm* expreffed in this *Table*, *b*. under
an *Iron-Pot* : it is fierce, and devours all other *Worms*,
even thofe that are wont to eat *Catterpillars* : I put him
(for the experiment fake,) with 4 other *Worms*, fuch as
are wont to devour *Catterpillars*, into a *China-Cup*, he
forthwith fnatched one of them up with his *Pincers*, and
fucked out all his Juice, and although the Yellow *Worms*
did much Strive to get away; yet this kept faft hold,
till he had killed them all.

Number. 120. *c*.

G. P. I.
Tab. 17.

Alfo again I noted about the Female of this *Species*
in the *Table*. *c*. that it was of a flow motion, and crept,
and refted by turns. It is wont to live in the darke un-

Q 3   der

der-ground, and alfo above ground in obfcure places.

It is nourifhed with *Worms* only that it finds, and devoures, it tears open the belly of the *Worm* it kills, and fucks the guts, and eats them, lifting up its head, and now and then fetling its feet upon the body of the *Worm*: I have obferved the guts of the Yellow *Worms*, which it had plucked out, to have the thicknes of the hair of a mans head, now when it is full, it fleeps 5 hours, wi hout any motion; and drawing its head almoft un-der its body, it lyes upon a fide.

And becaufe this *Beetle* is of a very flow motion, and lives not, but of other *Worms*, it has wings; thefe wings are as long as its body, and yet do not appear, or can be feen, unlefs it produce them; fo accuratly are they rolled up and folded, and thus hid, are not dirtied, and marred by creeping underground.

Section. 7.

OF

GRASSHOPPERS.

G. P. 2.
Tab. 40.

Number. 121.

Many years I defired to konw, whence *Grasfhoppers* were bred, and did ufe muc d ligence in the matter.

I have had 40. years experience, and do atteft the *Grasfhopper* depicted, in the 121 *Table*, yearly to be found, the moneths of *November*, and *Decmeber*. in our

Ifland

Ifland' of *Zeland*, about old *Lime-trees*, and to live for fome time of a certain humour they fuck from thefe trees.

For winter coming on, they all dye which come from this *Lime-tree*, and it feems proable, they muft defert thofe Trees; becaufe they afford them no longer nourifhment.

Whilft they are *Worms*, they live of and in the wood of the *Lime-tree*; but when they become *Grafhoppers*, they have need of other nourifhment then the leaves and humour of thefe trees.

The *Worm* depicted lying in the ftate of *Transmutation*, is not eafily found, unleffe when by the force of winds, the Trees are torn up by the roots, or when the wood is felled.

About that time this *Worm* light into my hands, having entered into the ftate of Tranfmutation and refting in it, fo that I cannot exactly tell the very time of its entry into the *Tranfmutation*, for it had placed it felf in the very interiour pith of the Tree, and I found it in the diffected wood the firft *January*.

As foon as I had got it, I placed it in its own earth, in a warm place near the fire. And the 31th. of the fame month it brought forth a *Grafhopper*, which caft its fhin firft, and then *changed* its colour. For 14 dayes I gave it *Sugar*, *Apples* and *Pairs*; but thefe not being its meat, it dyed.

Section.

Section. 8.
O F
FLYES.

G P. 1.
Tab. 53.

Number. 122.

The *Worm* of the 122 *Table*, was bred of the Putrid flesh of a certain Bird; which I kept under a glass, untill it did begin to be transformed, the 30 of *May*, and the 14th. of *June* it had the shape of a big Fly depicted.

When it first came forth of its worm, or egg, no wings did appear: but in the space of halfe an hour, it seemed bigger by halfe, and the wings were spread and smooth.

G. P. 1.
Tab. 54.

Number. 123.

The *Worm* of the 123d. *Table*, was bred of a *Flanders Pica*, dead, and corrupted.

Having sought a Place to hide it self in, after the manner of Insects, which *change*; it began its *change* the 12 of *June*; and, the 27 of the same month, it put on the forme of a *Fly*, depicted,

Number. 124.

G. P. 1.
Tab. 55.

The *Worm* of the 124. *Table.* had its originall from the corrupted brain of the *Water-Hen.*

When the time of Transformation comes, it boars a hole in fome wood, and hides it felf in it.

It boared open all the wooden boxes in which I endevoured to keep it ; fo that I was forft to keep it in a glafs.

It began to *change* the 12th. of *Auguft* : and it had the form of a *Fly,* the 26th. of the fame month, depicted.

Number. 125.

G. P. 1.
Tab. 69.

The *Worm* of the 125th. *Table,* had its originall from *Wheat-bran* putrefied ; which I kept mixt with water, till it grew fower, and began to putrefie.

From this Putrefaction came many *Little Wormes* ; which begun to take a new forme, altogether the 22d. day of *June* : and the 2d. of *July* in the fame year, they all *changed* into *Flyes* ; which lived without any food, till the 16th. of the fame moneth.

The following Flys *make a diftinct* Genus, *remarkeable for their* Quick *and* Steddie *flight.*

G. P. 1.
Tab. 2.

Number. 126.

The *worm* of the 126th. *Table:* is bred in *Jakes,* Privies, or fincks ; its *Long Taile* ferves, that it rolls not in creeping,

R.

ing, for that it has a round body without Legs; as it rolls it ballances it felf with its taile.

It compofed it felf for *change* the 26th. of *Auguft*, for which end, thefe *Worms* feek hiding places in the chinks, and clefts of old walls.

It remained in its *change* 17th. days, depicted in the *Table*, then came forth *a Bee*.

When this *Bee* was firft hatched, I thought it wanted wings; for in the place of them, I cou'd Animadvert nothing; but 2 white points like *Pin-heads*; but the Bee forthwith began to fpread thofe white fpots, with its hinder feet; and within 2 hours fitted them fo, that they were fit to fly with.

The nourifhment of this *Bee*, is a certain *fweet juice*; which it draws from the flowers of *Carduus Benedictus*, it alfo feeds on *Sugar*, and may be kept long alive with it.

It ufes little *Aliment*, for it lived 21 days without any food.

By the Bee, *is to be underftood a* Fly; *that is, a* Two winged *Infect; the* Bee kind *having ever* Four wings.

G. P. I. Tab. 4.

Number. 127.

The *Worm*, of the 127th. *Table*, is found in *Privies*; and is bred of *Mens Excrements*; it is a very flow creature.

It compofed it felf for *change* the 28th. of *September*: and remained therein to the 22d. of *October*.

And at that time was born a *Fly*, depicted in the *Table*.

Thefe *Flyes* are wont to abide about the places where they are bred, and I cou'd not keep it alive with *Sugar*.

(123)

Number. 128.

G. P. 1.
Tab. 4.

The *Worm* depicted in the 128*th*. *Table*, was bred of rotten dryed Fish called *Scate*.

Nature cannot be idle; the corruption of one is the generation of another; and from rottenneſſe, eſpecially in a hot and moiſt matter (which ſeems to be moſt apt for generation,) divers Animalls are produced.

I experienced this *Worm* the 26*th*. of *May*; to be *changed* into an egg; and a *Great Fly* to be bred of that egg the 21*ſt*. of *June*, depicted in the *Table*.

Number 129.

G. P. 1.
Tab. 50.

The *Worm* repreſented in the 129*th*. *Table*, is bred in the fruit of that *Shrub*, which in *Dutch* is called the *Speen appell*; and the virtues which are aſcribed to the fruit; are alſo of the *White worm* bred within it, to wit, for the *Piles*, and *burning Fevours*, and for theſe ends, the fruit is gathered about the 16*th*. of *October*, for the little *White worm* remains therein, to the middle of *June* of the year following; and at that time is turned into a *Fly*, making a hole to get out at.

Number. 130.

G. P. 1.
Tab. 51.

I had a mind to try, what wou'd become of the putrid and corrupted *Urine* of a man; I made a Funnell of paper, and ſo folded it, that no *Fly* or other *Animall* cou'd get into it : having infuſed into it oft times

R 2   humane

humane *Urine* ; I found fome *Worms* to be bred in the folds, where the feces ftayed.

I referved one of them, for the experiment fake, which compofed it felf for *change*, the firft of *March* : and the 14*th*. of the fame moneth , it had the forme of a *Fly*, depiᶜted in the *Table* : its head *was* Red, the body Black, and the hinder-parts Yellowifh.

G P. I.
Tab. 52.

Number. 131.

This *Worm* was bred of *Barley* flower ; which yet it did corrode into fmaller particles : this flower being moift, putrefied, and this *Worm* was bred form the putrefaᶜtion ; for the corruption of one, is wont to be the generation of another.

I cou'd not obferve this Worm to live ; but yet it began to be *changed* the 8*th*. of *Auguft*, and 23*d*. of the fame month it received the form of an *Oblong fly*.

G. P. I.
Tab. 73.

Number. 132.

The little Worm of the 132*th*. *Table*, was bred of *Rotten Cheefe*, mixed with Cummin ; with which alfo it is nourifhed, and kept alive.

Thefe *Worms* do not creep fo much, as (the body being gathered up) they fuddenly Spring, and Leap *Locuft-wife*.

Before they do compofe themfelves for *change*, they do quiefce 3 dayes.

This *Worm* began to *change* the 28*th*. of *November*, 1658. and the 10*th*. of *May* the year following, a little Fly appeared, whcih lived 8 days without food.

The

The *Worm*, the *Chryfalis* , and the Fly, are all depicted in the *Table*.

Number. 133.

G. P. I.
Tab. 41.

The *Worm* of the 133*d. Table*, is wont to be found
upon the leaves of the *Curran-bufh*, wh'ch are blafted :
under thofe leaves I have defcried to lie hid, a great
Number of little *Animalls* like *Pediculi*, the Worms of
this *Table* eat them.

For whilft thefe Worms do quiefce, without any
manifeft motion ; and that the *Animalls* do creep about
and upon them, they fting them with their beak, and eat
them.

It began to *change* the 15*th*. of *June*, and the 30*th*.
of the fame Month it appeared under the forme of a
Fly ; and fo in the fpace of a 11 Days a Fly is bred of
a Worm.

Thofe Pediculi *are very probably* Cimicces;

Number. 134.

G. P. I.
Tab. 47.

The *Worm* of the 134*th. Table*, feeds of little creatures, which they call Green Lice ; and do lick the
fat of Rofes ; fo that it is very manifeft, not the leaft
Animalls are fafe from enemies ; but the Lefs are food
for the greater.

This *Worm* do's ftick to things, and keeps it felf immoveable by the bigger and obtufe parts of the Body ;
but the acute and flender part, which is like a *Probofcis*, it toffes every way, and moves it, that it may

R 3   take

take some of the Forefaid little Creatures, and having
caught them , it lifts them up, left they fhould ftick or
adhere to any thing, and fucks them fo lifted up, untill
nothing remaine but a every thin Skin Wings and Feet.
Alfo thefe Worms feem to love *Ants*; for they are
often with them, but hurt them not.

It began to *Change* the 14*th.* of *July*, under the *Figure*
of an Egg : and the 21*ft*, of the fame Month, it received
the forme of a very long *Fly*.

This was a By-birth, *and an* Ichneumon-Wafp ; *the
true birth fhould have been a* Fly, *like the laft defcribed.*

G. P. 2.
Tab. II.

Number. 135.

There are many of thefe *Worms* to be found upon
the leaves of the *Elder* ; for they are bréd of the
Seed which the *Fly* their *Mother* layes upon them in
the Month of *June*, and Hatched by the heat of the
Sun.

They well know by inftinct (which is admirable)
that there will be food ready, as foon as their Young
ones are Hatched, this inftinct is *Analagous* to reafon,
and comes near it : So that there they lay their Eggs,
where they know the little Brood, as foon as Hatched,
will find meat; for the *Ants* do depofe a fertain fat
and fæcund humour, upon the extreme boughs of the
Elder, Apple, Cherry Pare-tree; and upon the leaves of
the *Curran-bufh*, from which little Creatures of a *Green*,
and fomtimes of a *Black colour*, are bred ; from
the

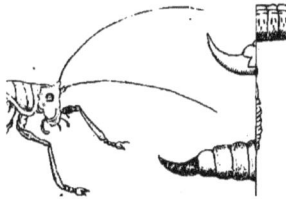

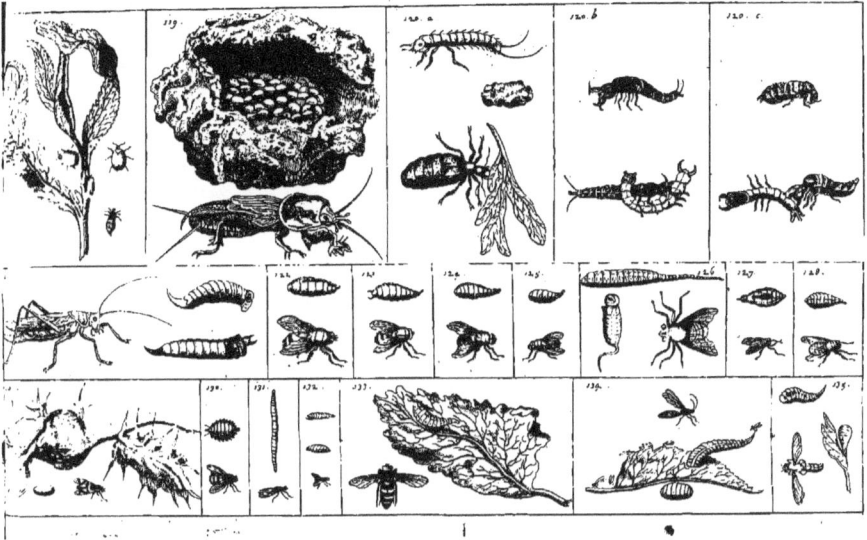

the hinder part of whose Body, an humour is wont to diftill, which these Creatures vehemently covert; as Fifhes Water, and greedily fup up.

And when these Flyes know, that these little creatures are bred, a future meat for their Young ones; they do injeft their Seed upon the leaves.

These little Creaturs have fix Feet, and long Winges, yet I never found them to fly; bnt are a food to thefe *Worms.*

These *Worms* know how to Counterfeit dead, and to lie amongft them; for that the innocent little Creatures creep over the Bodies of the Worms; which when the Worms feel, they take them with their fharp bills, and fnck them; lifting up their *Probofcis* like Hens, when they drinke.

These Worms feed of thefe little Creatures, and the Birds feed of the Flyes, they are *changed* into, and we of the Birds.

As foon as thefe little Creatures perceive the Worms, forthwith they fly them, and run away.

When this Worm has fed to its fullnefle for twenty four hours it is wont to reft and fleep; from Eight ith Morning till Nine the next Day, and when he again feels thefe little Creatures to run about him, he Eats as much of them as he has a mind, and now full again, it refts, and feems as Dead.

And the 12th. of *June,* this Worm affixed to a leaf did compofe it felf for *chunge,* as is Depifted.

But the 15th. of *July,* a *Fly* was Borne; which as foon as come forth, in halfe a quarter of an hour became as big again, and lived four Days without Food.

Thefe Little Creatures *are a numerous Brood of* Cimices.

Number. 136.

The 30th. of *August*, I took up a *Ripe Mushrome*; and puting it under a *Glasse*, I exposed it upon the Pavement to the Sun. The Day following, that Little *Mushrome* was full of little Black Worms; but the 11th. of *September*, the greatest part of it was turned into Black water, like Ink; except that part, which was called the Stoole, or Foot, being the Root.

In this impure Water I numbred 63 *worms*; which the Day following did betake themselves to *change*; but 7 Days after the 10th. of *September*, they were all turned to *Living Flyes*, very swift and nimble, with Red Heads, and Black Bodies; some of these were of an Elegant colour; and had on each side a little hammer, with which they did beat and excite themselves.

Some of these had Black heads; but one was of a different colour from the rest, and of a bigger bodie. &c.

You may keep these (if you please) long alive with a Sweet Liquor; I speak this by experience, as all things else; for I fed many of them from the 10th. of *September*, to the 24th. of *December* the year following; and I cou'd have easily kept them longer, if it had been to any purpose.

Number. 137.

We may very often see, upon the leaves of Oziers, and other Trees, certain *Little green creatures*, and

so

so tender-ſkined, that they are cruſhed with the leaſt touch, and the humour ſqueezed out of them is green.

Theſe are bred from a certain humour, which the Ants lay upon the ſaid leaves, and with the ſun, hatches into little creatures: they ſpoile the boughs they a re bred on; for they corrupt the wood, knitting upon it a net: under which lying-hid they grow up; and in the mean time do ſuck up all the moiſture; ſo that the boughs being deprived of that humour which is owing to them, do become lean, tender, and black barked; not like the reſt of the wood.

When they are new hatched, the *Ants* are ſeen to creep amongſt them, and as it were to cleanſe them.

The *Worm* depicted *Table* 13th. eats theſe little creatures, as ſoon, as they are grown up: yea it fills it ſelf with theſe, that it can hardly ſtir; and being ſo filled, it reſts quiet in the ſame place it laid it ſelf down in, to the next morning; and then, when the yeſter-dayes meat is digeſted, it begins to devour the remaining creatures, or ſeeks a new bough, better furniſhed, and on which more of theſe ſmall creatures ſit, and of them it fills its hungry belly, as before.

This *Worm* gave it ſelf to *change* the 10th of *June*, in that form and Poſture as is depicted, and it remained in that ſtate 20 dayes, ſo that the 29th. of *June*, a Fly came forth, which (wonderfull!) moſt ſuddenly came to its perfect bigneſſe; for within a quarter of an hour, it became as big again, as it was, when it was hatched, and as big again as the egg from which it broke out.

I knew not what to give it, that it wou'd taſt, and ſo it dyed the 5th. of *Auguſt*.

Number. 138.

The *Worm* of the 138th *Table*, is mostly upon the leaves of Elder ; for the most part it is on elegant whitish colour, and yellowish ; which afterwards becomes blackish.

It feeds of the little creatures, which are often to be found on the extream twigs of the Elder : these are the offspring of *Ants*; bred of a humid seed, layed by the *Ants* npon the extream boughs of the Elder, and, by the heat of the Sun, cherished and brought to maturitie ; that these little creatures might be bred of it. Also the *Ants* are continually at hand, to defend them against this *Worm. &c*:

It you wou'd know whence this *Worm* is bred, I answer that it is bred of such a Fly, as is depicted in the *Table*, which is wont to lay its seed upon the leaves of the Elder, well knowing these little creatures to be there, which may serve it for food, and from the seed of this Fly, this *Worm* cherished by the heat of the Sun, doth come.

When I had fed this for 23 dayes with its due food, it left eating, as though it wou'd have begun its *changes* but Two dayes after it put on the glorious colour before named ; which done, it again devoured the little creatures ; and last of all, the 26th *June* it did compose it self for *change*, and did remain in that state, to the 13 of *July*, on which day came forth a Fly, very like that of the 130th *Table* above, but a little bigger, and Yellow on the forepart, it lived fasting till the 17 of *July*.

R

Number. 139.

The *Worm* depicted in the 139*th Table*, is feldome found, for that it lyes in ftony, and moift grounds; and fometimes it is in Gardens, for the gardiners complaine that this *Worm* doth hurt their herbes not a little; nor is it eafily to be taken, for it knows how to hide it felf under ground, as foon, as it feels any thing to ftir.

It feeds on all forts of herbs, and eats their roots, it is very voracious, fo that filled it is twife as big, as when it is hungry.

It peeps out of the ground onely with the head; which it forthwith draws in again, as foon, as it doth perceive any noife; you muft take it with a Spade.

Having taken it, I put it in a glaffe, filled with earth, and did expofe it to the open aire; in a place moderately hot of the Sun; and I provided it with meat and drink; In this glafs it compofed it felf for *change*, in that pofture in which it is Painted; the 29*th* of *May* : and the 25*th* of *June* following, it bred a creature furnifhed with 2 wings, and 6 long Feet called by us when boyes, the *Tayler.*

Thefe *Tayler Flyes* are very *Leacherous*; there are double Number of males, for one female; for this you will experience with me, if you mind as I have done often, when the male Couples with the female moftly 5 or 6 males Fly about them, each of which ftay their courfe of coupling, and endevour to hinder the former; To whom it is given firft to couple, he remains affixt to her 2 dayes together; and the male, loofe, not very long after dyes; but the female lives.

And when the *seed* of the Female is ripe for laying, it infixes, its Taile in the ground, and layes its feed

in

in it; from which this mifcheivous *Worm* is bred, which is not fit to breed *Tayler-flyes* till it be Three Year old: if it were not for this Three Years unfit-neffe. to breed, and that there are more Males then Females; this *Worme* wou'd do great mifchiefe, in which, God's providence is much to be admired.

Thefe Flyes are thofe with very Long Legs, and which by our People are called Drummers. It is re-markable, that they lay Violet-coloured Eggs; all the Species af them, which I have yet feen, and have I feen at leaft Five forts with us in England.

Of the Originall of GNATS.

This is alfo a new Genus of Flyes

Number. 140.

G. P. 3. Tab. 10.   I have fought with great diligence, what was the generation of *Gnats*, and from what principle &c.

I Obferved *Gnats* to retire to *Cifterns* in great Numbers; and efpecially about the begining of *De-cember*; thither they carry their Seed, and after the fame manner, as I have often learnt by my own proper Experience; for they are wont to fit upon the furface of the Raine-watter, contained in *Cifterns*, the hinder part of the their Bodies being inclined down

downwards ; and to excerne their Seed, and that Seed
fo dejeted, doth forthwith finck to the bottom and
after a little time, is converted int Red, and (as it were)
Bloody little Infects.

Thefe *Little Bloody Worms* are nourifhed of certaine little Creatures fwimming in the Water, which I
have been wont to call Water Lice, becaufe they have
almoft the fame forme with the paper Lice, and they
ftay there 11 Months, and they do make themfelves
little Cells of their own juice, mixt with the particles
of Lime, within which they hide themfelves, againft
the extream cold.

But the time of the Naturall *change* being at hand,
they do convene in great numbers , and the heads of
all being joyned, the hinder parts of their Bodies,
are long much toffed , too and fro , and do excite
a great motion of Water , with which motion, I
have Obferved a great plenty of a certain *Tenacious
juice*, in which they compofe themfelves for undergo-
ing a *change*.

The 30th of *June*, and 13th of *July*,I Obferved many
Gnats to flie out of the forefaid Juice : The Males have
light Fethers on the Head, but not the Females ; but
do fwel with Eggs, nor yet can fo vehemently prick
our fkins as the males.

As *Gnats* do arife from bloody little *Worms*, fo they
do vehemently defire Humane blood ; for that end
they enter into our Honfes about Night , even by
the Chimneys, and other hidden places , if they find
not the Doors and Windows open, then they thruft
their Trunks, or Darts into the parts of the fkin :
Where they may find an open paffage to the Blood ;
which they fuck and are mightily delighted
in: They *Draw it* with very fubtle and *sharp
Darts*, made of as it were three fmall Hairs, contex-

S 3   con-

contexed; but hollow within like Pipes, and adapted to draw blood upwards.

I have fometimes Obferved, that a *Gnat* well filled with blood, not to recede from our Skin, untill a drop of Water, had been rejected out of her Bodie : Which I guefle to be the *Serum*; and in fo fhort a time to be feperated from the Blood, and rejected as an excrement.

When the *Gnats*, new bred, do afcend from the bottom of *Cifterns*, they are *white*, and do feem to want Feet, and Wings; but thefe do appear in the fpace of a quarter of an houre, and they are wont fo long to adhere to the fids and walls of the *Cifterns*; untill the wings are now enough expanded, are hard, and are fit for flight.

But fome will Object, how I cou'd fo accuratly Obferve, what *Gnats* do in the bottom of *Cifterns*; to whom I Anfwere, I Obferved all things in a Glafs Veffel made for that end, into which I put all thofe things, which are wont to be found in *cifterns*, to wit Lime, little Stones, Earth and Rain Water, to which I put the above defcribed Red *worms*. Alfo I was forced to place that Veffell in a cold place; alfo often to renew the Water, and to admit the frefh aire.

For experience taught me, thofe little *Worms*, for defect of Coolneffe and frefh and well tempered Aire, all foon to Die : Therefore by meanes of the pellucid Glaffe, I Daily and diligently obferved, what was Daily done in it, and what *changes* happened therein.

Section.

Section. 9.

OF

MILLEPEDS.

Number. 141.

GP. 2.
Tab. 36.

All this History *of the* Authors *is Founded upon a* mistake; *as thinking these* July, *or* Millipeds (*of all which He has given Us the Figured, Table* 141,) *were bred of the Thigh bone of a* Man; *when indeed, they were only lodged therein, in the depth of winter; the whole thing is impertinent and not worth the recitall.*

Number. 132.

G. P. 2.
Tab. 48.

Befides what I faid of *Mushrooms* Number 126. I faw alfo when the aforefaid *Worms* were changed into *Flyes*, and when I had placed the water, from whence they came, before the Sun; it was filled with exceeding little *Animalls*, which moved themfelves, and were therefore alive; which after I had taken out of the water, by the help of a Pin, and had view-ed ed them in a *Microfcope* ; I found them to be very little *Serpents*, or *Snakes*; and indeed, Multitudes of them did move fo fwiftly , and did mix themfelves fo together, that they cou'd not be counted.

I kept many of thefe two Years, and I found them marked in their Bodies ; fome with *Black*, fome with *Green*, and others with *Brown fpots* ; but the greateft of them which I kept, was of the Size and fhape, and variegated with *Brown fpots*, as is delineated in this 142 *Table* : It was very fwift and nimble, for as foon as I laid it down on the Ground , it crept into it, and I had loft it ; if I had not prefently dug it up, by the help of a little Spade, I had in my hand.

I kept this little *Snake* long in a Glafs, filled with Earth and water ; but at length through the great heat, and want of Water, he died, which I was very forry for having kept him fo long.

Section

O F

S P I D E R S.

Number. 143.

G. P. 2.
Tab. 49.

In the before mentioned Matter, from whence the *Flyes* and *Snakes* come, I saw also and found a certain kind of *Chryftalls* ; like in forme to small Sands, but but there grew out Feet to them, and by degrees they grew greater ; till in three Years space, they were full grown, and attained the forme of the *spider* Depicted in the 143 *Table.*

Of the *Mufhrooms*, growiug out of the Roots of the *Poplar Tree*, is wont to breed another kind of *Spider* ; of an oblong forme, a Yellow colour, and ftinking.

From *Rotten Mufhrooms*, another *Race of spiders* do's proceed, and they are *Reddifh*, there is another kind of *Mufhroom*, which rifes from the Earth, which firft of all appears fhut, but in a days time is open, and then it reprefents the forme of a *Mufhroom*, with a round difh, or little houfe, like the cup out of which the *Oake Acorns* are taken ; in this Cup is found Five round grains of feed, of the fize of Raddifh-feed, but pellucid, as *Chriftall* ; after thefe grains fall out of the cup on to the fuperficies, or into the chinks of the earth ; they are cherifhed by the heat of the

T   fun

fun, till they begin to live, after wards they get feet; and in three years time they attain to their full growth, all thefe things I tryed twice Two years together.

Thefe *Spiders* delight to be about the herbe *Balm*; and in Summer time they fhaddow themfelves under it; but in winter they ly hid in the Chinks of walls, or wooden fences.

The forementioned *Spider*, which has its originall of *Mufhrooms*, (which many efteem a delicate) has long feet, which I have exhibited to you, in its birth, progreffion, encreafe, and perfection, in this 143*d. Table.* where *Figure* 1 reprefents it, at it's firft appeareance, *Figure* 2 at the fpace of halfe a year : *Figure,* 3 at two years: *Figure* 4 Three years growth; when it has attained its full fize ; which it keeps till it dye, by a natural or a violent death.

Thefe *spiders* are rarely found on the fuperficies of of a fmooth wall, but commonly in the fiffures thereof; but fo as they have their feet exerted out of the Wall, and in my opinion to this purpofe; that if any humour come from the Wall, they may take it with their feet, for with this all their feet are embrewed, and they firft move their fore-feet to their Mouths, and then the reft; which they fucceffively brufh thorow their Mouths, by the help of their *Forcipes,* and fo lick off the *Nitrous humour,* fticking thereto, by which they live.

On the Day time thefe *Spiders* are wont to reft, unlefs you drive them away; but in the Night time they play together; for then they are often feen to take hold of one anothers feet, and fo to fall upon the ground, and when they play thus together, they do not hurt themfelves; for when they fall on the ground their long feet being ftreched out, they fall on them *like Cats* : As foon as they fall they prefently rife again, and ftradling with their Thighs, they afcend
the

the Wall with a Stilt-like motion : They often alfo
Fight fo, that they kill one another ; for they are very
tender, and are eafily killed ; and this is well feen in
them from an inftinct of nature ; therefore with their
legs contracted and affixed to both fides they fit, and
defend themfelves ; for then their bodies being de-
fended on every fide with their Feet, as Tents with
Pales, they fit fafe, prevailing againft the *Spiders*, that
affault them, with their own ftrength ; in this pofture
one of them cannot kill another at one affault ; what
do they then will you fay ? The more ftrong en-
tangle the Feet of the more weak with a wonderfull
dexterity, and hold them fo bound, like the fmall
tendrells you fee on vines, which wreath about and
encircle the branches, and embrace them ; and then
they hold their Feet fo enfnared with theirs, and
break them one after another ; nor do they ceafe, till
they have pulled off foure or five of them, this being
done, they leape upon them, and bite them, and fuck
the wound, and leave nothing, but an Exanguious
fkin, fuckt dry, and the reft of the Feet.

One of this kind of *Spiders* I kept long, with
Water of *Saltpeeter* and *Lime* mixt ; he was very fierce,
he alwayes overcame all the *Spiders* I offered him, although he found fome amongft them, which took him
up much time, and gave him fome danger of recovering his health, which he cou'd not overcome, but with
much difficulty ; at length he remained Victor of all,
viz. Thirty in number, which I gave him one after
another.

I placed before him two Pots ; in the one I gave
him Meat, in the other Drink ; but he put his Feet
into one, and the dry Carcaffes of the *Spiders* he had

Number. 144.

The *Worm* of the *144th Table*, is most worthy of admiration; which I took the Third Day of *Aprill* 1658. I never cou'd observe (though I was very diligent) whether it eat or drunk; neither cou'd I animadvert whether or noe it had any Eyes, or any aperture in its body; by which it might either receive nutriment, or eject excrement: it wants Feet; it always rests, immoveable, and rather rejoices in cold then heat; for I placed it in the Sun beams, and presently it sought for a skadow and shelter, and afterwards moved not.

sometimes it turned its body, so that it lay on its back, but presently contracting its body into a round, it kept its former habit, as is *Figured* in this *Table*.

I had this *Worme* alive with me at home, from the third of *Aprill* 1658, to the 28th of *August* 1659, without any food, and on that Day it dyed, afterwards (which is worthy to be observed,) there lived always near this *Worm* three *White Animalls*, lesse then common *sand*; which always stuck on the belly, or back of the *Worm*, and after this manner they lived Nine Months without any food that I observed, at length, two of these *Animalls* on the Head, and the third on the Back of the *Worm*, dyed.

F I N I S.

136.

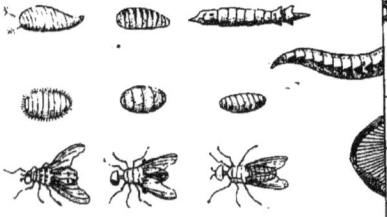

140.

143.

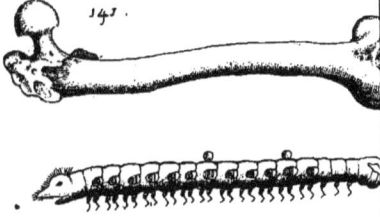

144.

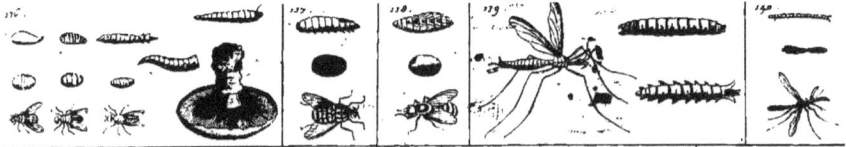

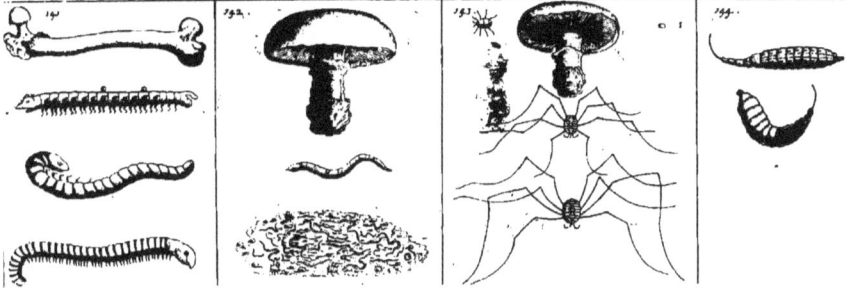

www.ingramcontent.com/pod-product-compliance
Lightning Source LLC
Chambersburg PA
CBHW021710210326
41599CB00013B/1596